**Differential Space,
Quantum Systems,
and Prediction**

Norbert Wiener
Armand Siegel
Bayard Rankin
William Ted Martin

Differential Space, Quantum Systems, and Prediction

The M.I.T. Press
Massachusetts Institute of Technology
Cambridge, Massachusetts, and London, England

Preface

The idea of writing the present book was crystallized in the summer of 1956, following two years of work in a probability seminar at the Massachusetts Institute of Technology. Between the summer of 1956 and the present, some severe recasting of the original idea took place. It was felt from the first, however, that there was a natural association between the Brownian motion process and differential space, the prediction theory based on stationary processes, and the quantum concepts based on a statistical-mechanical, differential-space approach. Despite other changes, this merging of subject matter has survived. It was also felt that the diverse interests of the authors provided a broader base for a book than would be possible with any one author alone. In particular, the diverse subject matter just mentioned could be developed in some detail by having the authors concentrate in their fields of specialty, when writing was concerned, and by having them take the role of critics, when questions between coauthors and questions concerning the book as a whole arose.

It should be especially mentioned that the material of this joint work would undoubtedly never have been viewed as a unity without the guiding interest of the senior author, the late Norbert Wiener. The seminar first mentioned was devoted largely to tracing his contributions in the historical development of Brownian motion theory and statistical prediction and his contributions in the many applications, including recent work on quantum theory.

The desire on the authors' part for their work to appear as one book was matched with their willingness to yield on some editorial points for the best interest of a readable result. For that reason one member accepted the responsibilities of editor.

In anticipating an audience for the range of subjects covered, it was inevitable to require a knowledge of some rather well advanced areas of mathematics. While it is hoped that any interested scholar, upon reading through a portion of each chapter, will be encouraged to learn more of this area of thought, a detailed understanding of all the chapters will certainly require the fundamentals of measure theory, measure-theoretic probability, and some notions of Hilbert space. In the chapter on matrix factorization and prediction, it is assumed that the reader is familiar with the single-series linear prediction, though the outline of the single-series solution appears in the introduction. The notation of Dirac is used in the chapter on quantum theory, but it is explained in full.

The reader will recognize in the chapters entitled "The Brownian Motion Process and Differential Space" and "Matrix Factorization and Prediction" many of the ideas that have appeared in the articles and books of Norbert Wiener. Of this material, the detailed construction of the Brownian motion process has been completely reworked and rewritten for this book. The matrix factorization concerns the positive Hermitian case and is a reprinting of the first four sections of Wiener's 1955 article in *Commentarii Mathematici Helvetici*. The reprinting is supplemented here with extensive footnotes. The prediction theory, while using Wiener's successive projection method that has appeared before in the two-dimensional case, provides for the first time a scalar solution to the n-dimensional prediction problem.

The differential space, matrix factorization, and multiple prediction theories given here are in the spirit of Wiener's original presentations, though all of these basic theories have experienced many changes of form since Wiener's pioneering contributions. Though much elegance, generality, efficiency of notation, and simplicity could have been gained by presenting these ideas in their secondary form (that form worked out after the original conceptions have been labored through), the authors felt that the student and the research worker would not be benefited. For Wiener, the working form was frequently the constructional form. While his work may constitute in this respect a special case, quite different from the norm, the authors felt that to strip his presentation of much of the original construction would lead the reader further from, not closer to, original ideas.

The chapter entitled "Integration in Differential Space" is devoted to three main ideas: integration with respect to sample functions of the Brownian motion process, the Fourier-Hermite development of nonlinear functionals, and the evaluation of various integrals involving the Brownian motion process. The first of these ideas was originally defined and studied by Wiener and provides a necessary background for the following two. The Fourier-Hermite development and the later evaluation of various integrals are, with the exception of some work of Kac and Lévy on integration involving the Brownian motion process, an exposition of the journal articles of William T. Martin in collaboration with R. H. Cameron, and appear in book form for the first time.

The final chapter on quantum theory is the work of Armand Siegel and Norbert Wiener. It presents in detail the researches of these authors that have appeared only in the journals. Briefly stated, this chapter approaches the fundamental Bohr-Einstein controversy on the physical interpretation of quantum mechanics with the tools of Brownian motion theory. A part of the program of this chapter is to show how ensembles of completely described systems obeying a postulated dynamics can be so constructed as to have the same statistical properties that are expressed by a given quantum-mechanical wave function. An Appendix that supplements the work of this chapter has been provided by Professor Robert L. Warnock.

During the actual writing, the work was shared among authors according to chapter and section. Armand Siegel did the writing of the chapter on quantum theory. William T. Martin wrote the two concerning Brownian motion and integration in differential space. Norbert Wiener wrote the sections on matrix factorization. Bayard Rankin wrote the introduction and the section on prediction.

Some important acknowledgments are due. *Commentarii Mathematici Helvetici* has very kindly permitted the reprinting of the article "On the Factorization of Matrices" by Norbert Wiener. The Case Institute Research Fund and a research fund granted to the

Massachusetts Institute of Technology Department of Mathematics by the Ramo-Wooldridge Corporation both greatly facilitated the writing of this book by awarding one of the authors (B. R.) a grant and a fellowship. During the two years of the seminar, the Office of Naval Research supported the work of the same author. The Guggenheim Foundation, by awarding a Guggenheim fellowship to Armand Siegel during the year 1957–1958, helped make the contributions by that author possible.

The M.I.T. Press has from the first encouraged the writing and has, with helpful advice, seen the manuscript through many forms.

BAYARD RANKIN, *Editor*

Case Institute of Technology
February 1966

Contents

Differential Space,
Quantum Systems,
and Prediction

I

Introduction

I. Objectives

Most of the ideas of measure-theoretic probability theory, both physical and mathematical, are in the literature of the early part of this century. It is significant that at the same time that measure theory was being formulated by Borel, Lebesgue, and other mathematicians in France, the concept of the statistical ensemble in statistical mechanics was being propounded independently by Gibbs in America. The two concepts were not clearly united into a probability theory before Wiener's 1923 work in differential space, although one can see in retrospect all the probabilistic interpretation of measure theory in the work of Gibbs and all the measure-theoretic foundations for a probability theory in the work of Borel and Lebesgue. The ergodic properties of systems, like the statistical concepts, are now most frequently studied in the setting of measure-theoretic probability theory, although they originated with the ergodic hypothesis of Boltzmann as desirable properties of a physical system. The occurrence of ergodic properties in the areas of Brownian motion theory, prediction of stationary processes, and quantum theory can be traced to the setting of classical Hamiltonian mechanics from which each of these theories grew. These facts have more than mere historical importance for the organization of the present material. Throughout, we wish to keep in view the original merging of mathematical and physical concepts into what is now understood to be measure-theoretic probability. The measure-theoretic way of thinking about probabilistic ideas began with physical concepts and

I

now finds natural application in those physical sciences (particularly, quantum mechanics) that we wish to discuss.

2. Measure theory and probability

As ingredients for a (finite) measure theory, one has, to use modern terminology, the three entities of a (finite) *measure space*, $(\Omega, \mathfrak{B}, \mu)$. They satisfy the following conditions, where B^c is the logical complement of B, and $B_1 \cup B_2 \cup \cdots$ is the logical union of $B_1, B_2, \cdots$:

    i. Ω is a space of points.
   ii. $\mathfrak{B}$ is a collection of subsets of Ω that form a σ-field. That is:
       (a) $\Omega \in \mathfrak{B}$.
       (b) If $B \in \mathfrak{B}$, then $B^c \in \mathfrak{B}$.
       (c) If $B_1, B_2, \cdots \in \mathfrak{B}$, then $B_1 \cup B_2 \cup \cdots \in \mathfrak{B}$.
  iii. μ is a measure on $\mathfrak{B}$. That is, μ is a real-valued, nonnegative function of the sets in $\mathfrak{B}$ such that
       (a) $\mu(\Omega)$ is finite.
       (b) $\mu(B_1 \cup B_2 \cup \cdots) = \sum \mu(B_k)$, whenever $B_1, B_2, \cdots$, are pairwise disjoint.
  iv. $\mathfrak{B}$ contains all subsets of sets $N \in \mathfrak{B}$ for which $\mu(N) = 0$.

The sets $B \in \mathfrak{B}$ are called *measurable sets*. If Ω is chosen to be a finite segment of the real line (or if it is chosen to be the real line and iii(a) is weakened to σ-finite) and if $\mathfrak{B}$ includes all intervals in the segment, then the conditions i to iv define in postulational form the generality of Lebesgue's measure theory. As explained in Section 4 of this chapter, however, Lebesgue's own approach to measure theory was not so much postulational as constructional. In a natural way it included not only an existence proof for a nontrivial measure space (into which the above four postulates give us no insight) but an immediate fulfillment of condition iv. Lebesgue's work was done in 1902. In 1898 Borel had preceded him in measure theory with the generality as defined by conditions i to iii above, but Borel's work was not adapted to include condition iv in a natural way. The extremely subtle condition iv identifies the class of *Lebesgue measurable sets* as compared to the smaller class of *Borel measurable sets*.

If $\mu(\Omega) = 1$, the conditions just stated for a measure space define a *probability space*. If Ω is chosen to be the interval [0, 1], and $\mathfrak{B}$ includes all intervals in [0, 1], the measure space is of sufficient generality to serve (either directly or through a Borel mapping) for all of the measure-theoretic investigations undertaken in this book.

Birkhoff's work [1931] on the ergodic theorem, that depended on von Neumann's delayed publication [1932a] on the mean ergodic theorem, opened a period of rapid growth. Between 1931 and 1933 the full power of Boltzmann's ergodic hypothesis (suitably weakened as will be described in Section 4) was explored. A point transformation T of Ω in a (finite) measure space $(\Omega, \mathfrak{B}, \mu)$ is the ingredient that corresponds to a dynamics in Birkhoff's theorem. A one-to-one point transformation T, carrying points of Ω into Ω, is called *measurable* and *measure-preserving*, if it satisfies the following conditions:

   v. $TB \in \mathfrak{B}$ and $T^{-1}B \in \mathfrak{B}$, whenever $B \in \mathfrak{B}$.
   vi. $\mu(TB) = \mu(T^{-1}B) = \mu(B)$, whenever $B \in \mathfrak{B}$.

The ingredients in Birkhoff's theorem that correspond to the phase functions of statistical mechanics are the real-valued functions $f(\omega)$ defined on Ω, that are measurable with respect to $\mathfrak{B}$ and integrable in the Lebesgue sense. A function $f(\omega)$ is called *measurable with respect to $\mathfrak{B}$*, if for any real number t

$$\{\omega : f(\omega) < t\} \in \mathfrak{B}.$$

A measurable function is *invariant* to transformations of Ω effected by the transformation T, if $f(\omega) = f(T\omega)$ for almost all ω.

Using these concepts, Birkhoff's theorem states that for any measurable function $f(\omega)$ that is integrable in the Lebesgue sense, and for any one-to-one point transformation T defined on Ω, that is measurable and measure-preserving, the limit

$$\lim_{N \to \infty} \frac{1}{N} \sum_{1}^{N} f(T^k\omega)$$

exists as a measurable, integrable, and invariant function, except for a set of points ω that depends on $f(\omega)$ and has measure zero (see Wiener [1939]).

Needless to say, because of its broad implications, Birkhoff's theorem had value for many parts of mathematics. Its importance for statistical mechanics comes mainly from a corollary that results by assuming an added property of T. The added property is that of metric transitivity: A one-to-one point transformation T of Ω that is measurable and measure-preserving is called *metrically transitive*, or *ergodic*, if for any $B \in \mathfrak{B}$

$$TB = B \quad \text{implies} \quad \mu(B) = 1 \quad \text{or} \quad \mu(B) = 0.$$

(The space Ω is sometimes called *metrically transitive* or *metrically indecomposable* with respect to a transformation T, if T is metrically transitive.) It follows immediately from Birkhoff's theorem that, if T is metrically transitive, then the limit function of the theorem is just the Lebesgue integral of $f(\omega)$ over Ω.

In physical language, Birkhoff's theorem (or more correctly, the corollary to the theorem) gave precise conditions under which the average behavior in the time evolution of a physical system could be related to the phase average of that behavior (behavior being expressed through a given phase function). The mathematical setting for Birkhoff's ergodic theorem, with the help of certain added concepts, was the same that Wiener had used in the study of differential space. Thus, the tools that Birkhoff used in 1931 to formulate and prove his ergodic theorem about the time evolution of a mechanical system, like those used by Wiener in 1923 to define and investigate the properties of the Brownian motion process, were the already well-established tools of Lebesgue measure theory.

By 1933, the measure-theoretic components of a probability theory were sufficiently clear that Kolmogorov [1933] could set down, with almost no reference to a specific historical background in physics, the axioms of measure-theoretic probability theory. In his fundamental paper, he sounded with finality the acceptance of measure theory as the proper setting within which to prove the classical theorems of probability theory and to explore the properties of stochastic processes. After the pattern of definitions set down by Kolmogorov, a real-valued, finite function defined on Ω that is measurable with respect to $\mathfrak{B}$ is called a *random variable*. If this variable is integrable in the Lebesgue sense, its integral is called its *mathematical expectation.*

Because the probabilistic techniques of this book are entirely measure-theoretic, it is well to comment upon the limitations that measure-theoretic techniques place upon the scope of probabilistic studies.

For purposes of illustration, we will introduce, with accompanying examples, the concept of a probability functional space (Rankin [1960]). This space is consistent with, but more general than, the probability space that has just been defined. All the work of this book could be carried out under the postulates of a probability functional space, although it is generally unknown when the stronger requirements of a probability space are necessary for the results obtained in the chapters to follow. The ensuing postulates are more general than those of probability spaces by virtue of the weakening of the familiar complete additivity, or continuity, postulate for an additive set function (condition iii*b*). One might look upon the definitions to follow as a system for defining a class of random variables possessing minimum closure properties without calling upon the complete additivity postulate, that is, without calling upon the whole mechanism of Lebesgue theory. It is interesting to recall that the continuity axiom is the only axiom introduced by Kolmogorov in his basic paper of 1933, of which he said, "It is almost impossible to elucidate its empirical meaning." It is because of its extreme importance in facilitating the mathematics that the axiom is kept throughout most of probability theory and throughout this book.

Let $S \equiv \{a\}$ be a collection of elements a, $[0, 1]^N$ the closed N-dimensional unit cube, $\bar{R}^N$ the extended N-dimensional real space, and H^N a subset of $\bar{R}^N$. We write $\mathfrak{C}(H^N)$ for the class of all functions defined on H^N to $[0, 1]$ and continuous on H^N. We will also write $\bar{R}$ for $\bar{R}^1$.

Let $f_1, f_2, \cdots, f_N$ be N functions, each defined on S to $\bar{R}$. If H^N is the range of the vector function $(f_1, f_2, \cdots, f_N)$, a finite composition $c \circ (f_1, f_2, \cdots, f_N)$ of the vector function $(f_1, f_2, \cdots, f_N)$ with the continuous function c is defined for each $c \in \mathfrak{C}(G^N)$, $H^N \subset G^N$, by $c \circ (f_1, f_2, \cdots, f_N)(a) = c[f_1(a), f_2(a), \cdots, f_N(a)]$.

A class $\mathfrak{F}$ of functions, each defined on S to $[0, 1]$, is said *to be closed under all finite compositions with continuous functions*, if $c \circ (f_1, f_2, \cdots, f_N) \in \mathfrak{F}$ for all finite collections $f_1, f_2, \cdots, f_N$ chosen from $\mathfrak{F}$, and for all $c \in \mathfrak{C}([0, 1]^N)$.

The following postulate will be essential for defining a probability functional space.

Postulate 1 $\mathfrak{F}$ *is a nonempty class of functions defined on S to* [0, 1] *and closed under all finite compositions with continuous functions.*

Let us denote by "$\mathfrak{L}(\mathfrak{F})$" the smallest linear space containing a class $\mathfrak{F}$ of functions defined on S to [0, 1], that is, the space of complex-valued functions with domain S such that

   vii. $f \in \mathfrak{L}(\mathfrak{F})$, if $f \in \mathfrak{F}$.
   viii. $k_1 g_1 + k_2 g_2 \in \mathfrak{L}(\mathfrak{F})$, if $g_1, g_2 \in \mathfrak{L}(\mathfrak{F})$ and k_1, k_2 are finite complex numbers.
   ix. $\mathfrak{L}(\mathfrak{F}) \subset \mathfrak{L}'(\mathfrak{F})$ if $\mathfrak{L}'(\mathfrak{F})$ is any other space satisfying conditions vii and viii.

Let $\mathfrak{F}$ be a class of functions defined on S to [0, 1] that satisfies Postulate 1. A *probability functional* associated with the class $\mathfrak{F}$ is a complex-valued functional E defined on $\mathfrak{L}(\mathfrak{F})$ and satisfying the following four properties:

   x. $0 \le Ef \le 1$,   for $f \in \mathfrak{F}$.
   xi. $EI_S = 1$,   for $I_S \equiv 1$.
   xii. $E(k_1 f_1 + k_2 f_2) = k_1 Ef_1 + k_2 Ef_2$, for $f_1, f_2 \in \mathfrak{L}(\mathfrak{F})$ and k_1, k_2 complex.
   xiii.[1] For any vector function $(f_1, f_2, \cdots, f_N)$ with f_k in $\mathfrak{F}$, $k = 1, 2, \cdots, N$, and for any uniformly bounded sequence of functions $c_1, c_2, \cdots$, defined and continuous on R^N, such that $c_n \to 0$ pointwise on R^N, it follows that as $n \to \infty$,

$$Ec_n \circ (f_1, f_2, \cdots, f_N) \to 0.$$

[1] This property was not included in the original definition of a probability functional space (Rankin [1960]). On the other hand, in the earlier reference, all theorems involving characteristic functions and distribution functions of random variables were obtained under the assumption that the probability functional space was Riemann.

A *probability functional space* is the triplet $(S, \mathfrak{F}, E)$ in which S is a space of points, $\mathfrak{F}$ is a class of functions defined on S to $[0, 1]$ that satisfies Postulate 1, and E is a probability functional associated with $\mathfrak{F}$. A *random variable* defined on the probability functional space $(S, \mathfrak{F}, E)$ is a real, finite function X whose domain is S and for which

$$\mathfrak{F}_X = \{c \circ (X): c \in \mathfrak{C}(\bar{R})\} \subset \mathfrak{F}.$$

Example 1 *As an example of a probability functional space, let S be the unit interval $[0, 1]$, $\mathfrak{F}$ the class of continuous functions defined on and to $[0, 1]$, and Ef the Riemann integral of $f \in \mathfrak{F}$. In this case the class of random variables X defined on $(S, \mathfrak{F}, E)$ coincides with the class of real continuous functions defined on $[0, 1]$.*

Example 2 *As another example let S be the unit interval $[0, 1]$, $\mathfrak{F}$ be the class of Riemann integrable functions defined on and to $[0, 1]$, and Ef be the Riemann integral of $f \in \mathfrak{F}$. In this case the class of random variables X defined on $(S, \mathfrak{F}, E)$ coincides with the class of real finite functions defined on S which are continuous almost everywhere.*

We know from Lusin's Theorem that the class of real finite functions defined on $[0, 1]$ which are almost continuous coincide with a class of random variables (measurable functions) defined on a real probability space $([0, 1], \mathfrak{B}, P)$. From this fact it is clear how the random variables X of this example compare with random variables defined on $([0, 1], \mathfrak{B}, P)$.

Example 3 *As another example, let $(S, \mathfrak{B}, P)$ be a probability space, $\mathfrak{F}$ be the class of measurable functions defined on S to $[0, 1]$, and Ef be the integral or mathematical expectation of f on the space $(S, \mathfrak{B}, P)$. In this case the class of random variables defined on $(S, \mathfrak{F}, E)$ coincides with the class of random variables defined on $(S, \mathfrak{B}, P)$, and if $A \in \mathfrak{B}$, then $I_A \in \mathfrak{F}$ and $EI_A = PA$, where I_A is the indicator of the set A.*

Example 4 *Let $S = (a_1, a_2, a_3, \cdots)$ be a uniformly dense sequence in the closed unit interval $[0, 1]$ (see Weyl [1916]). Let $\mathfrak{F}$ be the class of Riemann integrable functions defined on and to $[0, 1]$ and restricted to S. For $f \in \mathfrak{F}$, let*

$$Ef = \lim \frac{1}{N} \sum_{1}^{N} f(a_k).$$

It can be shown that Ef, as defined here, equals the Riemann integral of any Riemann integrable function on [0, 1] whose restriction to S is f. E can immediately be extended to $\mathfrak{L}(\mathfrak{F})$, and (S, $\mathfrak{F}$, E) is a probability functional space. The class of random variables defined on (S, $\mathfrak{F}$, E) can be obtained by restricting to S the class of real finite functions on [0, 1], which are continuous almost everywhere.

An important thing displayed in the first two examples is that the class of random variables defined on a probability functional space may coincide with a subset of the class of random variables defined on a probability space. The third example shows that the concept of random variables on a probability functional space, though a generalization, is not inconsistent with the measure-theoretic concept. The fourth example suggests in what sense probability functional spaces permit a generalization of probability spaces. Within the context of Example 4, it is possible to select a sequence of functions $f_1, f_2, \cdots$, each belonging to $\mathfrak{F}$, which permit the following two things:

$$f_n \to 0, \qquad \text{monotonely on } S; \qquad Ef_n \to 1.$$

Thus, the fourth example allows an evident infringement of the monotone convergence of Lebesgue theory. Under the restrictions of Example 4, however, many of the results of measure-theoretic probabilistic analysis remain valid. In particular, infinite numbers of statistically independent random variables possessing any pre-assigned distribution functions can be defined and many of the attendant limit theorems for independent random variables are meaningfully preserved.

The deeper questions, such as whether constructions of the kind that appear in the next chapters could be accomplished under the restrictions of Example 4, appear to be largely unexplored. It is known, however (Rankin [1960, 1966a]), that certain computable constructions for random variables, as well as certain probabilistic treatments of an individual system (as contrasted with an ensemble of systems, or population), are possible *only* in the absence of complete additivity. Further remarks concerning the probabilistic treatment of individual systems—in particular the quantum mechanical treatment of the individual electron—will be made in Section 7. As will be remarked in that place, an interpretation of the behavior of the electron may be possible *only* in the absence of additivity. Thus, measure theory does appear to place some limitations on probabilistic studies, though only indirectly on the kinds of questions considered in this book.

3. Metric transitivity and mixing

As indicated in Section 1, the work of Gibbs contributed physical motivation for the measure-theoretic approach to probability. Gibbs' contributions, as well as the basic contributions of Einstein, Smoluchowski, and others concerning the Brownian motion phenomenon, led to the first rigorously defined stochastic process, namely, the Brownian motion process of Wiener [1923]. Thus there was in the early part of this century a new sophistication toward the post-Renaissance statistical approach to chance that had begun with the work of Pascal, Huygens, Jacob Bernoulli, de Moivre, and others. Toward the end of the last century, on the other hand, much of the work of the physicist Boltzmann, particularly his ergodic hypothesis (or the more acceptable quasi-ergodic hypothesis), was in reality an attempt to evade the statistical point of view in physics. This being the case, the so-called "ergodic approach," historically predating the "ensemble approach" of Gibbs, is an additional physical origin of modern probability theory, one which, in fact, foreshadowed a non-Gibbsian element (Rankin [1966a]). In spite of Boltzmann's approach, his ergodic hypothesis has suggested a basic concept in modern measure-theoretic probability theory.

Just how one can refer to the ergodic approach as being non-statistical, in view of the eminently fundamental place that ergodic theory has in measure-theoretic probability theory today, can be understood briefly as follows. The ergodic approach as originally conceived in statistical mechanics is not represented in present-day statistical and ergodic theory. Boltzmann's ergodic hypothesis states that for a conservative mechanical system of fixed total energy the time evolution of the system describes, in terms of physical states successively occupied, an individual trajectory that constitutes the entirety of all possible physical states of the system. On the other hand, the measure-theoretic refinements of this idea (aside from the refinements of the quasi-ergodic hypothesis) involve the measure-theoretic concept of metric transitivity. The latter concept is weaker than Boltzmann's original ergodic hypothesis because it allows more than one trajectory as possible evolutionary histories of the system and requires merely that the set of all possible trajectories shall not be decomposable into proper measurable subsets. The concept of metric transitivity made it possible to transfer properties suggested by Boltzmann's ergodic hypothesis into measure theory without actually transferring the hypothesis itself.

The importance of ergodic properties, as they occur in mechanics, is that they provide a link between averages that extend over an individual trajectory and averages that extend over all possible trajectories of a physical system. From the point of view of probability theory, the ergodic properties of mechanics (as postulated by assuming that the Hamiltonian dynamics is metrically transitive) provide a physically realizable model for the existence of constant limits of the type

$$\int_\Omega f(\omega) \, d\omega.$$

Metric transitivity, a concept of physical origin, is thus at the foundations of the logic that relates probability, in the theoretical sense, with the type of averaging processes that go into the application of probability theory. It is also true that a further property of measure-preserving transformations, which is a highly plausible property of certain physical systems, provides a logical building block for the concept of statistical independence in probability theory. This property is called *mixing* (see Hopf [1937] and Halmos [1956]).

Using Birkhoff's ergodic theorem, it is easy to show that a measure-preserving transformation T defined on the (finite) measure space $(\Omega, \mathfrak{B}, \mu)$ is metrically transitive if and only if

$$\lim_{N \to \infty} \frac{1}{N} \sum_1^N \mu(A \cap T^k B) = \mu(A)\mu(B) \tag{1}$$

for every $A, B \in \mathfrak{B}$.

It is also easy to show that if $\lim \mu(A \cap T^k B)$ exists for every A, $B \in \mathfrak{B}$, then Equation 1 implies

$$\lim_{N \to \infty} \mu(A \cap T^k B) = \mu(A)\mu(B), \qquad (2)$$

for every $A, B \in \mathfrak{B}$. Condition 2 is called *strong mixing*, and its relation to statistical independence is obvious.

Some remarks on the mixing property of the Brownian motion will be made in Section 5.

4. The constructional method in mathematics

It was suggested in Section 2 that Lebesgue's approach to measure theory was more constructional than postulational. It will also be shown that Wiener's approach to differential space involves a construction of measure from basic principles. In this section we wish to clarify these statements and to point the way to a valuable technique that is used in much of the work in this book.

Lebesgue's approach to measure theory, like that of Borel, began with the measure of open and closed sets of points on the real line. (To simplify these brief remarks, we will speak only of sets contained in an interval of finite length.) The measure of an open set was defined to be the sum (or limit of the sum) of the lengths of a finite (or countable) number of open, nonoverlapping intervals whose logical union coincided with the open set. The measure of a closed set was defined to be the difference of the measures of two open sets whose logical difference coincides with the closed set. This method of defining the measure of open and closed sets was consistent with the idea of length of an interval, in that the measure and the length of an interval were the same. Moreover, the concept of measure, extended to the class of open and closed sets, displayed the characteristic properties of length, such as additivity. Lebesgue, in effect, made use of a theorem on the construction of measure for open and closed sets. The theorem states that the measure of either an open or a closed set is the upper bound of the measures of closed sets contained in it and also the lower bound of the measures of open sets containing it. He defined the full class of sets any one of which had the property of possessing a common value for the upper bound of the measures of closed sets contained in it and for the lower bound of the measures of the open sets containing it. This was the class of *Lebesgue measurable sets* and the common value of the upper and lower bounds was the *Lebesgue measure.*

An interval of finite length, the Lebesgue measurable sets within this interval, and the Lebesgue measure constitute an important measure space and satisfy all the conditions i to iv stated in postulational form in Section 2. It should be noticed that even the condition iv, a condition not at all anticipated from the properties of the class of intervals, is a natural property of the class of Lebesgue measurable sets.

Lebesgue's measure theory, in the first instance, is confined to sets of points in the real line. Once it is stated in this limited context, however, it is a relatively easy job to abstract its defining characteristics (as was done in Section 2, properties i to iv) and to show that the essential features are not limited in application to sets of points in the real line. There is a positive advantage to stating the defining characteristics of a theory in postulational form, but it is not a substitute for constructing a theory by starting with specialized logical concepts and progressing through steps of higher and higher logical type. The postulational approach, however elegant and useful, is either a second stage of refinement in mathematical thought, or, allowing it to be an inspired approach unanticipated by mathematical experience, it is in need of an existence proof that demonstrates its nontrivial character.

Generally speaking, mathematics benefits from the contributions made through both the constructional and the postulational approaches. The latter is by far more strongly emphasized in modern mathematics, however, and a student frequently fails to see that the two approaches can, in many instances, be essentially identical. It is important for a full appreciation of much of the material of this book to recognize the constructional approach. The following are the defining features:[2] First a logical simulacrum of a given system is

[2] Suggested by a private communication of Norbert Wiener, June 5, 1959. A larger quotation from the same communication is the following:

In modern mathematics there is an alternative to the postulational method which, as a matter of fact, is one which has always appealed to me. Instead

obtained. Next, by manipulating constructions in a higher logical type, one obtains systems agreeing with postulates except perhaps in one or two details.

Examples of the constructional approach are to be found throughout the book. An obvious example is found in the use of the unit interval as a space from which to build up a probability theory. Only the most delicate uses of analysis will reveal the unthought-of degree of generality that can be built up from this apparently restricted setting. The advantages, both intuitional and technical, that are gained from working in the unit interval far outweigh the slightly greater generality that would be achieved by postulating an abstract probability space at the outset. As we have already stated in Section 2, the probability space based on the unit interval is sufficiently general to serve for all of the measure-theoretic investigations undertaken in this book. When the advantages are sufficiently great, the unit interval notation is used explicitly.

Another example is to be found in Wiener's approach to the Brownian motion process (Chapter 2). The method used there for constructing a measure on a space of functions is to establish a correspondence between a certain subclass of functions (those that will be the sample functions of the process) and points on the unit interval. A correspondence is made at the same time between subsets of functions and subsets of points in the unit interval. Thus measure in the highly complicated function space can be built up from the

---

of starting from a certain stated set of conditions and then determining extrinsically whether there are any systems satisfying these conditions, one very often has the alternative of building logical constructs of a higher type and thus disposing both of the existence proof and the postulational treatment of the system at one blow. This method has a considerable history to it. While Huntington was a postulationalist, there are several pieces of work of his in which the constructional method is very clearly indicated, as for example in some of his work in geometry. An even more thorough devotee of these methods was A. N. Whitehead. . . . I myself have done a considerable amount of work in this direction, as you will see from my early logical papers. As a matter of fact, Gödel and, following him, Turing, have made essential use of the fact that, once you have a logical simulacrum of a given system, by manipulating constructions in a higher type you can obtain systems agreeing with postulates except perhaps in one or two details. . . . I have always been struck with the constructional method as a varied alternative or supplement to the postulational.

rudiments of Lebesgue measure in the unit interval. Moreover, integration of functionals of the Brownian motion process can be built up with the techniques of Lebesgue integration. Wiener's method, like Lebesgue's, incorporates an existence proof and builds an important property, the separability of the Brownian motion process, naturally into the system, just as Lebesgue's built into the measure the property of *completeness* (property iv of Section 2). In the postulational approaches to the Brownian motion processes (see remark at the end of Section 2, Chapter 2), a separability postulate must be somewhat artificially invoked in order to ensure that a stochastic process can be manipulated with sufficient freedom (Doob [1953], p. 51). The generality that is gained in the postulational approach to stochastic processes by having the freedom to use or not to use the separability postulate can be compared to the generality that is gained in measure theory with regard to property iv of Section 2. Some further discussion of Wiener's Brownian motion construction will be given in the next section.

A valuable technique in the study of stationary time series, $\cdots, X_{-1}(\alpha), X_0(\alpha), X_1(\alpha), \cdots$, is to represent the series with a slightly specialized construction involving the unit interval, a one-to-one point transformation T of the unit interval, that is measurable and measure-preserving, and a measurable function f that is of class L_2. That is, we write, for $n = 0, \pm 1, \pm 2, \cdots, 0 \leq \alpha \leq 1: X_n(\alpha) = f(T^n\alpha)$. In fact, this is the representation that was chosen by Wiener for much of his work in prediction theory and in the factorization of matrices. The specialization obtained does not prevent complete exploration of all the analytical subtleties of single series and multiple prediction. At the same time, explicit calculational devices can be used that are close to the intuition and close to the original physical motivation based in the Hamiltonian dynamics (see Section 2). Such a constructional method will be used throughout Chapter 4.

In regard to the prediction technique of Chapter 4, it should also be pointed out that an explicit representation of the predictor in terms of the past data of the original time series is used. Once again, this is a constructional device characteristic of Wiener's methods, and analytic conditions that justify such a representation were originally indicated by Wiener [1949b] and were later studied in greater detail by Akutowicz [1957] and Wiener and Masani [1958]. Thus, by slightly limiting the generality of the prediction, explicit calculational formulas are obtained. At the heart of the matrix factorization of Chapter 4 is a successive projection technique, which permits the explicit predictor representation to be carried into higher dimensions, with the aid of an added assumption.

The last chapter makes pivotal use of one or another of two algorithms that appear at first glance to be of a special and artificial character. These are the so-called dichotomic and polychotomic algorithms. The algorithms are used for the assignment of realized values of a physical observable (in the quantum mechanical sense) to points in differential space. This amounts to formulating physical observables as functions of the sample functions of the Brownian motion process. The chapter goes on to prove that the distribution function of any physical observable, as constructed in this apparently ad hoc fashion, agrees with that of quantum mechanics. Comments on the advantages of the construction, beyond its mere agreement with quantum mechanics, conclude the chapter.

5. The Brownian motion process

As we mentioned in Section 2, Wiener's study of Brownian motion in 1923 was a significant step toward the measure-theoretic analysis of probabilistic phenomena (and thus a step toward the measure-theoretic probability theory). To a large extent, the qualitative features of the physical Brownian motion that had been discovered by Robert Brown in 1827 were understood by the time Wiener approached the subject. Prior to 1923, however, there was little indication of the surprising mathematical properties that would result from a rigorous mathematical description of the physical phenomenon, and in particular there was little indication of the extremely valuable implications that were in store for many branches of science and engineering. Though Wiener's measure was motivated in a special way (in terms of a mathematical description of the Brownian motion), it served as a prototype for much of the later work on measure and integration in function space. His method of inducing a measure space through a mapping procedure suggests an elegant and useful contemporary mathematical tool, a generalization of the measure-preserving transformation that we have defined, which, in its generalized form, maps one space onto another.

Although Chapters 2 and 3 will contain a detailed study of the Brownian motion process, starting from the constructional definition as originally given by Wiener, in this section we will indicate the essential features of the construction as well as the basic properties of the process.

What is needed is a collection $\{X(t, \alpha), -\infty < t < \infty\}$ of random variables such that for (almost) all α the function $x(t) = X(t, \alpha)$ has the properties characteristic of the wandering of a particle suspended in a fluid and moving under the impacts of the molecules of the fluid. The particular choice of the range of the time parameter t, here taken to be $-\infty < t < \infty$, is somewhat immaterial, mathematically, and will be identified elsewhere with the unit interval. The same remark holds for the space of points α, and, following the remarks of Section 2, we will choose that space to be the unit interval. Thus the probability space on which the random variables are defined is the finite Lebesgue measure space $([0, 1], \mathfrak{B}, \mu)$.

By properties characteristic of a Brownian motion particle, we understand the following:

Writing $\Delta X_t = X(t + \tau, \alpha) - X(t, \alpha), \tau > 0,$

   i. *For arbitrary $\tau > 0$, and arbitrary t, ΔX_t is a Gaussian (normal) random variable with mean zero and variance τ.*

  ii. *For arbitrary $\tau > 0$, and for $s < t - \tau$, ΔX_t and ΔX_s are statistically independent.*

 iii. *For $t = 0$, $X(t, \alpha) = 0$ for almost all α.*

Property iii is a simplifying property. Einstein [1905, 1906] investigated the consequences of properties i and ii for the physical Brownian motion, under the assumption that τ can be selected as large compared to the time intervals between molecular collisions, and provided certain other reasonable physical assumptions are met. Properties i and ii, as stated for arbitrary τ, are mathematical properties abstractly assumed for the collection $\{X(t, \alpha), -\infty < t < \infty\}$, and they permit the use of limiting operations and other tools of analysis in the study of this entity, the Brownian motion process.

The primary step that Wiener made in his mathematical investigation was to prove by construction the existence of a collection of random variables with the properties i to iii. This done, he was able to show the surprising results of his construction: (a) That for (almost) all α, the sample function $x(t) = X(t, \alpha)$ is continuous, (b) that for (almost) all α, the sample function is of unbounded variation, and (c) that for (almost) all α the sample function is nondifferentiable.

Another important property of the Brownian motion process is obtained by defining a one-parameter group of measure-preserving point transformations, T^λ, $-\infty < \lambda < \infty$, that act upon the unit interval. The group is defined through time-translates of increments of the process. For all λ, for all τ, and for all t,

$$\Delta X_{t+\lambda}(\alpha) = \Delta X_t(T^\lambda \alpha). \tag{3}$$

Wiener proved (Paley and Wiener [1934]) that the transformation defined by Equation 3 is metrically transitive.

From the properties i to iii of the Brownian motion process, it is clear that

$$P\{X(t, \alpha) < a\sqrt{t}\} = \frac{1}{\sqrt{2\pi}} \int_{-\infty}^{a} e^{-u^2/2} \, du. \tag{4}$$

Statements such as Equation 4, as well as like statements that give the distributions of various functionals of the Brownian motion process, have important interpretations in many areas such as diffusion theory, heat conduction, communication theory, and stellar dynamics. A study of various functionals is found in Chapter 3.

A brief introduction to Wiener's method of construction will now be given, anticipating greater detail later in the text. In this construction, we identify the time axis with the unit interval.

Wiener's method for constructing a measure on function space (in particular for constructing a collection of random variables with properties i to iii) is to establish a correspondence between subsets of functions and subsets of points in the unit interval. We consider the class C of all real-valued functions x defined on the unit interval $\{t: 0 \le t \le 1\}$ and vanishing at the origin. Within this class C, for each positive integer n, we define $(2^n)^{2^n}$ subsets Q as follows:

For the (2^n)-tuple $(m_1, m_2, \cdots, m_{2^n})$, constructed by selecting each number m_j from among the 2^n numbers

$$-2^{n-1}, -2^{n-1} + 1, \cdots, 2^{n-1} - 2, 2^{n-1} - 1,$$

let $Q = Q(n; m_1, m_2, \cdots, m_{2^n})$ be composed of the class of all functions x in C that satisfy at each of the dyadic points of the form $a = j/2^n, j = 1, 2, \cdots, 2^n$, the following inequality:

$$\tan (m_j \pi/2^n) < f(j/2^n) \le \tan [(m_j + 1)\pi/2^n], \text{ when } m_j > -2^{n-1},$$
$$\tan (m_j \pi/2^n) \le f(j/2^n) \le \tan [(m_j + 1)\pi/2^n], \text{ when } m_j = -2^{n-1}.$$

It is characteristic of these subsets Q that a subset of the form $Q(n; m_1, m_2, \cdots, m_{2^n})$ is a logical union of a finite number of subsets of the form $Q(n + 1; m_1, m_2, \cdots, m_{2^{n+1}})$.

We can write

Case A $C = \cup Q(n; m_1, m_2, \cdots, m_{2^n}),$

where the union extends over all 2^n-tuples with n fixed.

Case B $Q(n; m_1, m_2, \cdots, m_{2^n}) = \cup Q(n + 1; m_1, m_2, \cdots, m_{2^{n+1}}),$

where the union extends over all 2^{n+1}-tuples, with $n+1$ fixed, which are contained in $Q(n; m_1, m_2, \cdots, m_{2^n})$.

We now associate positive numbers with each of the Q's, the number $\mu(Q)$ being associated with the subset Q in such a way that the following three properties are satisfied:

1. $1 = \sum \mu\{Q(n; m_1, m_2, \cdots, m_{2^n})\}$, in case A.
2. $\mu\{Q(n; m_1, m_2, \cdots, m_{2^n})\} = \sum \mu\{Q(n + 1; m_1, m_2, \cdots, m_{2^{n+1}})\}$, in case B.
3. For $\delta > 0$ and for λ in $0 \le \lambda < \frac{1}{2}$, we can delete an appropriate countable collection of subsets Q' for which $\sum \mu(Q') < \delta$, such that the remaining subset will contain only functions of C that satisfy the condition

$$|x(t_1) - x(t_2)| \le h|t_1 - t_2|^\lambda$$

for some $h = h(\delta) > 0$, and for all dyadic points in the unit interval.

For Wiener, the fact that the preceding properties 1, 2, and 3 can be satisfied was a consequence of a special selection for the numbers $\mu(Q)$. Motivated by the prior work of Bachelier and Einstein on Brownian motion, Wiener made the identification

$$\mu(Q) = (2\pi)^{-p/2}\left[\prod_1^p (t_j - t_{j-1})\right]^{-\frac{1}{2}}$$

$$\times \int_{r_1}^{s_1}\cdots\int_{r_p}^{s_p} \exp\left[-\frac{1}{2}\sum_1^p \frac{(u_j - u_{j-1})^2}{t_j - t_{j-1}}\right] du_1\cdots du_p, \quad (5)$$

where

$$Q = Q(n; m_1, m_2, \cdots, m_{2^n}), \qquad r_j = \tan(m_j\pi/2^n),$$
$$s_j = \tan[(m_j + 1)\pi/2^n], \qquad p = 2^n, \qquad t_j = j/2^n,$$
$$t_0 = 0, \qquad u_0 = 0.$$

For the present, this special identification need not concern us.

We first use properties 1 and 2, and for each $n = 1, 2, \cdots$ we partition the unit interval into $(2^n)^{2^n}$ subintervals, in such a way that for fixed n the subintervals have lengths $\mu\{Q(n; m_1, m_2, \cdots, m_{2^n})\}$; and successive partitionings are accomplished by partitioning each interval of length $\mu\{Q(n; m_1, m_2, \cdots, m_{2^n})\}$ into subintervals of lengths $\mu\{Q(n + 1; m_1, \cdots, m_{2^{n+1}})\}$ that satisfy property 2.

We next identify subsets Q with subintervals in the unit interval in such a way that subset Q is associated with a subinterval of length $\mu(Q)$, and, if $Q(n + 1; m_1, m_2, \cdots, m_{2^{n+1}})$ is contained within $Q(n; m_1, m_2, \cdots, m_{2^n})$, then the corresponding subintervals are contained within one another. Now assume identification 5.

Thus, every point in the unit interval, except for a set of Lebesgue measure zero, is uniquely characterized by and uniquely characterizes an infinite sequence of subintervals successively contained in one another and with length tending to zero. It is also true that every point in the unit interval, except for a set of Lebesgue measure zero, is uniquely characterized by and characterizes an infinite sequence of subsets Q in C successively contained in one another and with associated numbers $\mu(Q)$ tending to zero.

Now, for a given point in the unit interval which is not one of the end points of one of the constructed subintervals, we might ask whether the sequence of subsets Q associated with this point determines a unique function. If x_1 and x_2 are two functions common to the sequence of subsets in question, then $x_1(t) = x_2(t)$ for all dyadic points in the unit interval. Thus, we can conclude that if there is a continuous function common to the sequence of subsets, it is unique.

Using property 3, we reason as follows. To each function f satisfying the inequality of property 3 for all dyadic points, we can associate a unique continuous function X, namely, the function defined by $X(t) = x(t)$ for a dyadic point t, and $X(t) = \lim x(t_n)$ otherwise, where the limit is taken along dyadic points approaching t as a limit. This associated continuous function X satisfies the inequality of property 3 for every pair of points in the unit interval. We can now use Ascoli's theorem. The functions X satisfying the inequality of property 3 form an equicontinuous class and are uniformly bounded. Thus any sequence of such functions has a uniformly convergent subsequence, whose limit is, itself, a continuous function satisfying the same inequality. Using property 3, we see that by removing a set of arbitrarily small measure from the unit interval, every point α that remains is characterized by a uniquely determined continuous function $X(t, \alpha)$ that satisfies the inequality of property 3. It is this final function of the pair (t, α) that constitutes the Brownian motion process.

An intuitive understanding of the interesting properties of the Brownian motion process can be obtained immediately from another construction that will now be outlined. Although it is closely related, this construction, which would lead to a completely rigorous definition of the Brownian motion process if carried out in detail, is distinct from Wiener's construction. In this construction, it is convenient to make the time axis coincident with the whole real line.

As in proofs for the existence of an infinite number of statistically independent random variables, consider the measure space $([0, 1]^\infty, \mathfrak{B}^\infty, \mu^\infty)$ composed of the countably infinite dimensional hypercube, the corresponding product Borel field, and the product measure. Denoting arbitrary points on the orthogonal axes of the hypercube by $\beta_1, \beta_{-1}, \beta_2, \beta_{-2}, \cdots$, an arbitrary point in the hypercube may be written

$$\beta = (\beta_1, \beta_{-1}, \beta_2, \beta_{-2}, \cdots). \tag{6}$$

Thus, to obtain an infinite number of statistically independent random variables, one need only define on the infinite dimensional measure space a sequence of finite measurable functions that depend on successive orthogonal axes:

$$Y_k(\beta) = Z(\beta_k); \qquad k = 1, -1, 2, -2, \cdots. \qquad (7)$$

For arbitrary nonnegative integer n, we are free to select each of the statistically independent random variables with distribution function

$$F_n(a) = \frac{1}{\sqrt{2\pi 2^{-n}}} \int_{-\infty}^{a} e^{-u^2/2 \cdot 2^{-n}} \, du. \qquad (8)$$

With the random variables of definition 7, which depend on n, we can define the collection

$$\{X(t, \beta), \qquad t = 0, \pm 2^{-n}, \pm 2 \cdot 2^{-n}, \cdots\} \qquad (9)$$

as follows:

$$X(m2^{-n}, \beta) = \frac{m}{|m|} \sum_{k=0}^{m} Y_k(\beta); \qquad Y_0(\beta) = 0, \qquad m = 0, \pm 1, \pm 2, \cdots. \qquad (10)$$

It may be observed that the collection 9 possesses all of the properties i to iii required of the Brownian motion, provided t and τ that appear in the statements of these properties are chosen to be of the form $m2^{-n}$, $m = 0, \pm 1, \pm 2, \cdots$, n arbitrary, nonnegative but fixed.

There exists a measure-preserving mapping U of the countably infinite dimensional hypercube onto the unit interval, which is one-to-one (up to a set of measure zero), with which the random variables 9 can be transformed according to

$$X(m2^{-n}, \alpha) = X(m2^{-n}, U^{-1}\alpha); \qquad 0 < \alpha < 1. \qquad (11)$$

The random variables on the left of Equation 11 are thus defined on the measure space $([0, 1], \mathfrak{B}, \mu)$. The explicit construction of such a mapping U has been studied by Wiener in a number of references (for example, Paley and Wiener [1934]).

The random variables of Equation 11 can be further restricted by suitably choosing the variables $Y_k(\beta)$ depending on n, and by suitably choosing the mapping $U = U_n$, depending on n, so that, for dyadic

t, X depends only on the pair (t, α) and has properties i to iii required of the Brownian motion, provided t and τ that appear in the statements of those properties are chosen to be dyadic.

The final stage of the construction extends the domain of definition of $X(t, \alpha)$ from dyadic t to all real t. As in Wiener's construction, this extension must be accomplished with the use of certain equicontinuity properties. From the present approach, it can be shown by direct computation that for λ fixed in $0 \leq \lambda < \frac{1}{2}$, for h and L any positive numbers, and for random variables chosen from the collection 11, that the following inequality holds:

$$\mu\{\alpha: |X[(m + 1)2^{-n}, \alpha] - X(m2^{-n}, \alpha)| > h2^{-n\lambda}, -L < m2^{-n} < L\}$$
$$< w(h, L), \qquad (12)$$

where $\lim_{h \to \infty} w(h, L) = 0$. Thus, for $\epsilon > 0$, $L > 0$, and $0 \leq \lambda < \frac{1}{2}$, there is an h, such that upon excepting a set of points α of measure less than ϵ, the following condition holds for fixed α and for any integers m and n, such that $-L < m2^{-n} < L$:

$$|X[(m + 1)2^{-n}, \alpha] - X(m2^{-n}, \alpha)| \leq h2^{-n\lambda}. \qquad (13)$$

With the use of Ascoli's theorem, as cited above, inequalities of the type 13, stated for a succession of real numbers ϵ converging to zero, identify for each fixed α outside a set of measure zero a unique continuous function $X(t, \alpha)$, $-L < t < L$.

In this way, we can obtain the Brownian motion process for which the continuity of almost all sample functions is evident. The nondifferentiability and nonbounded variation properties are examined in a similar way.

From this approach, one can also immediately see the metrically transitive and strong mixing properties of the transformation defined in Equation 3. First, we must consider the rotation of the orthogonal axes of the countably infinite dimensional hypercube. By a rotation, we understand a transformation R operating on points given in Equation 6 such that

$$R\beta = (\beta_2, \beta_1, \beta_3, \beta_{-1}, \cdots). \qquad (14)$$

This is clearly a measurable and measure-preserving transformation of the hypercube. The transformation is also strongly mixing, as we can prove by examining its effect on sets S in the hypercube defined by a finite number of inequalities of the type, $a_k < \beta_k < b_k$, and generalizing the result. For such a set S, it is obvious from the properties of the rotation transformation and the product measure that for some integer N_0 and for all integers $N > N_0$

$$\mu(S \cap R^N S) = [\mu(S)]^2. \tag{15}$$

Thus, upon generalizing, the strong mixing property given by Equation 2 is obtained, and, from the remarks of Section 3, metric transitivity of the rotation transformation also follows.

The rotation transformation now generates, through the measure-preserving mappings U_n, a one-parameter group of strongly mixing transformations on the unit interval. The group is obtained by defining,

$$T^\lambda(\alpha) = U_n R^m U_n^{-1}(\alpha), \tag{16}$$

for $\lambda = m2^{-n}$, and by extending the definition to all real λ. The strong mixing property of the Brownian motion transformation then follows by identifying the transformations defined in Equations 3 and 16.

The type of analysis used in this discussion makes clear the intimate association of the statistically independent increments ΔX_t of the Brownian motion process and the orthogonal axes of a space: thus, the origin of Wiener's use of the term *differential space* to describe the Brownian motion. An intuitive grasp of this association will become especially helpful in Chapter 5 on the differential space theory of quantum systems.

Regardless of the irregular (continuous but nondifferentiable) character of the sample functions of the Brownian motion process, Wiener was able to define integrals of the type

$$\int_{-\infty}^{\infty} \phi(t)\, dX(t, \alpha), \tag{17}$$

where $\phi \in L_2$. Though Wiener defined these integrals by use of a technique that has since become associated with Schwartz's theory of distributions, these integrals may also be defined, as is done in Chapter 3, by starting with step functions that are dense in L_2 and by applying the Riesz-Fischer theorem.

In Chapter 3 it is also shown that for functions f chosen from a suitable class, and for functions $\phi_k(t)$ belonging to L_2 over $-\infty < t < \infty$, the class of random variables of the form

$$f_q(\alpha) = f\left[\int_{-\infty}^{\infty} \phi_1(t)\, dX(t, \alpha), \cdots, \int_{-\infty}^{\infty} \phi_q(t)\, dX(t, \alpha)\right] \quad (18)$$

for some integer q, are everywhere dense in the L_2-space over the unit interval. This result is taken from the work of Cameron and Martin [1947] and suggests one of their main results which is also proved, namely, that it is possible to build up a complete orthonormal set based on functions of the type in Equation 18 with which any function in L_2-space can be approximated.

Applications of this result are in the direction of nonlinear analysis, though they are not discussed in this book.

6. Prediction theory

In this section we will present the linear prediction of single series stationary processes that results by specializing the multiple prediction of Chapter 4 to the one-dimensional case. Thus, the present remarks are not intended as a complete theory, since in Chapter 4 some prior knowledge of the single series case is assumed. Without the complication of matrix factorization, however, the present remarks may indicate the basic techniques of prediction theory that will be used in this book.

Both Wiener [1949a] and Kolmogorov [1941a, 1941b] contributed to the early development of prediction theory. The presentation given in this section is probably best described as a merging of their methods, though it follows closely a little-known article of Wiener [1949b].

The probability space that underlies this work will be $(A, \mathcal{B}, \mu)$, where $A \equiv [0, 1]$ is the unit interval, $\mathcal{B}$ is the corresponding Borel field, and where the probability measure μ corresponds to the ordinary Lebesgue measure.

Let $f(\alpha)$ be a complex, measurable function belonging to class L_2, and let T be a one-to-one point transformation (measurable and measure-preserving) carrying the unit interval onto itself. The time series under consideration will be the complex stochastic process $\{f(T^n\alpha); \quad n = \cdots, -1, 0, 1, \cdots\}$ with the index n, the power of the transformation, corresponding to increment steps on the time axis. For convenience we will take $n = 0$ as the present and associate negative powers of T with the past.

The *prediction* of the complex random variable $f(T^m\alpha)$, $m > 0$, will be some function of the present and past values of the just described stochastic process which is close in some sense to the value of $f(T^m\alpha)$. The criterion of "closeness" is conventionally defined in terms of mean-square deviation from the true value, and for the sake of simplicity only the closest among all *linear* combinations of the present and past are considered. Thus, we will assume that the linear prediction of $f(T^m\alpha)$ will be some linear combination of the form

$$L^{(m)}(\alpha) = \sum_{n=0}^{\infty} f(T^{-n}\alpha)P_n^{(m)}, \qquad (19)$$

with convergence defined in the L_2-metric, and the "best" among all these linear forms will be that which minimizes

$$\int_0^1 |f(T^m\alpha) - L^{(m)}(\alpha)|^2 \, d\alpha. \qquad (20)$$

The minimum value of the integral 20, namely $[\sigma^{(m)}]^2$, is the prediction error. Analytical conditions under which the predictor can be represented in the form appearing in Equation 19 were indicated for the first time by Wiener [1949b] and were studied in more detail by Akutowicz [1957].

For the present solution, we make the following assumption:

i. *The projection of $f(\alpha)$ on the closed linear manifold generated by the set of functions,*

$$f(T^{-n}\alpha), \, f(T^{-n-1}\alpha), \cdots,$$

converges in the mean to zero as n approaches infinity.

Assumption i implies that, with $m > 0$, the prediction error given by the minimum value of the integral 20 is positive. The condition for a positive prediction error is often expressed analytically as the convergence of a logarithmic integral. The form of the integral is stated in Chapter 4 where the spectral distribution function for the stochastic process is also defined. In terms of these analytical concepts, assumption i states that the spectral distribution function is absolutely continuous and the logarithmic integral converges. The exact formulation of assumption i in terms of the spectral distribution will not be needed here.

A reference to the special nature of the stationary process considered here was made in Section 4. For further questions concerning generality, see Doob [1953, pp. 452–457].

The objective of our discussion is to express $f(T^m\alpha)$, for $m > 0$, as the sum of two parts:

$$f(T^m\alpha) = F^{(m)}(\alpha) + G^{(m)}(\alpha), \tag{21}$$

where $F^{(m)}(\alpha)$ is linearly dependent on the present and past or, in other words, expressible in the form 19 and where $G^{(m)}(\alpha)$ is orthogonal to the present and past.

In symbols

$$\int_0^1 G^{(m)}(\alpha)\overline{f(T^{-n}\alpha)}\, d\alpha = 0; \qquad n = 0, 1, 2, \cdots.$$

If we succeed in this, $F^{(m)}(\alpha)$ will be the unique closest point to $f(T^m\alpha)$ in the manifold generated by the present and past, since $f(T^m\alpha) - F^{(m)}(\alpha) = G^{(m)}(\alpha)$ is orthogonal to that manifold. In other words, $F^{(m)}(\alpha)$ will be the best linear prediction of $f(T^m\alpha)$ as already defined.

Throughout the remainder of this section we will use the conventional inner product notation. Let $u(\alpha)$, $v(\alpha)$ be any measurable functions belonging to L_2 over the interval $[0, 1]$. We write

$$(u, v) = \int_0^1 u(\alpha)\overline{v(\alpha)}\, d\alpha, \qquad \text{and} \qquad (uT, v) = (u \circ T, v).$$

We will also use the projection notation, where $Pf(\alpha)$ is the projection of $f(\alpha)$ onto the closed linear manifold in L_2-space that is orthogonal to the past of $f(\alpha)$ (or, more precisely, orthogonal to the closed linear manifold that is generated by the past of $f(\alpha)$).

Let us define the *innovation* function for the process

$$h(\alpha) = \frac{Pf(\alpha)}{\|Pf\|}, \qquad \text{where} \qquad \|Pf\|^2 = (Pf, Pf). \tag{22}$$

This function is linearly dependent on the present and past of $f(\alpha)$, and represents the part of $f(\alpha)$ that is orthogonal to its own past. Therefore, with assumption i the set $\{h(T^{-n}\alpha); \; n = 0, 1, 2, \cdots\}$ represents a complete and orthonormal set for the manifold generated by the present and past of $f(\alpha)$. We have

$$(hT^{-n}, hT^{-m}) = \begin{cases} 1, & n = m, \\ 0, & n \neq m; \end{cases}$$

$$(f, hT^n) = 0, \qquad n > 0; \tag{23}$$

$$f(\alpha) = \sum_{n=0}^{\infty} h(T^{-n}\alpha)(f, hT^{-n}), \qquad \text{in the } L_2\text{-metric.} \tag{24}$$

That is to say, $f(T^m\alpha)$, $m > 0$, can be expanded in the desired form given in Equation 21 where (in the sense of the L_2-metric)

$$F^{(m)}(\alpha) = \sum_{n=m}^{\infty} h(T^{m-n}\alpha)(f, hT^{-n}), \tag{25}$$

$$G^{(m)}(\alpha) = \sum_{n=0}^{m-1} h(T^{m-n}\alpha)(f, hT^{-n}). \tag{26}$$

Expression 25 is one form of the prediction, though it does not clarify in what way the prediction is a linear combination of the past of the original time series. From the integral 20 and the remarks following it, as well as Equations 21, 23, and 26, we see that the prediction error can be expressed in the form

$$[\sigma^{(m)}]^2 = \sum_{n=0}^{m-1} |(f, hT^{-n})|^2. \tag{27}$$

It remains to determine the coefficients (f, hT^{-n}) appearing in Equations 25 and 26 and the coefficients $P_n^{(m)}$ appearing in Equation 19 in terms of the basic data available from the original time series. The basic data derivable from the original time series are the covariances (fT^{-n}, f).

Let us further assume that (with convergence defined in the sense of the L_2-metric) we can analyze the innovation function as an explicit linear combination of the present and the past of the original stochastic process

$$h(\alpha) = \sum_{n=0}^{\infty} f(T^{-n}\alpha)a_n. \tag{28}$$

This assumption can be made meaningful in an analytic sense by reference once again to the work of Wiener [1949b] and Akutowicz [1957]. From the expression 28 and the properties 23 of the innovation function given in that equation, it is seen that the coefficients a_n are deducible from the coefficients (f, hT^{-n}) through the recurrence relations

$$(f, h)a_0 = 1,$$

$$\sum_{n=0}^{v} (f, hT^{n-v})a_n = 0; \qquad v > 0. \quad (29)$$

Thus, by combining Equations 25 and 28, under the assumptions we have stated, it follows that the prediction coefficients $P_n^{(m)}$ are also deducible from the coefficients (f, hT^{-n}). That is, with convergence defined in the sense of the L_2-metric, the predictor becomes

$$F^{(m)}(\alpha) = \sum_{n=0}^{\infty} f(T^{-n}\alpha) \sum_{v=0}^{n} a_{n-v}(f, hT^{-v-m}); \qquad m > 0. \quad (30)$$

With these remarks, we have reduced the single series problem of determining the linear least-squares predictor and the prediction error to the problem of finding the coefficients (f, hT^{-n}) from the covariances (fT^{-n}, f). The solution to this last step has been extensively discussed in the literature and will be summarized for the single series case in Chapter 4, Section 5, where it will also be shown that the pattern of reasoning that we have used here generalizes to the multiple series case.

7. Quantum systems

From the rudiments of measure-theoretic probability theory (or the more general probability functional space theory of Section 2) there can be defined, for each random variable X, a distribution function F (a nonnegative, left-continuous, nondecreasing function of a real variable) that relates to the mathematical expectation of X, through the representation

$$E(X) = \int_{-\infty}^{\infty} x \, dF(x). \quad (31)$$

Distribution functions also enter in the calculus of quantum theory, though these are not customarily related to well-defined measurable functions on measure spaces. Instead, for each physical observable with corresponding Hermitian operator $\mathbf{R}$, with eigenspectrum $\{r\}$, and with corresponding resolution of the identity $\mathbf{E}_R$, one writes, for state vector ψ,

$$F_R(r) = (\psi, \mathbf{E}_R(r)\psi), \tag{32}$$

where F_R is a distribution function that is related to quantum mechanical expectation through the identity

$$(\psi, \mathbf{R}\psi) = \int_{-\infty}^{\infty} r \, dF_R(r). \tag{33}$$

In the study of quantum systems that composes Chapter 5 (deriving from the work of Wiener and Siegel [1955]), as well as in a number of other sources (for example, Rankin [1965]), it is of particular interest *to identify a distribution function of the type F_R with a distribution function F of a measurable function defined on some measure space of points.* Though this is a purely mathematical formulation, the problem thus identified is of basic physical interest. On more careful examination, as will be presented in Chapter 5, the quest for an underlying measure space in the theory of quantum systems relates directly to the quest for an appropriate physical interpretation of a single quantum mechanical system. As is well known, from Bohr's point of view, the wave function ψ can represent a single system—an individual. In fact, this point of view has become closely associated with the successful resolution of the wave-particle duality paradox. On the other hand, as will be indicated in Section 1 of the final chapter, it may also be appropriate to regard ψ as representing an *ensemble* of individuals. More precisely, it will be assumed permissible to adapt to quantum mechanics the approach of Gibbs, namely to consider all statistical predictions of a quantum mechanical system as obtainable from an ensemble of dynamically precisely defined systems.

It becomes evident from these remarks that the problem that we formulated in mathematical terms immediately after Equation 33 relates to the long-standing controversy concerning the proper physical interpretation of quantum mechanics. One point of view in this controversy, the more generally accepted one, has become associated with the name of Niels Bohr, the other point of view with Albert Einstein. In the writings of von Neumann [1932b], Bohm [1952, 1953, 1957], Wiener and Siegel [1953, 1955], and others, the term "hidden variable theory" is used to identify theories that are consistent with Einstein's interpretation of quantum mechanics. To quote Wiener and Siegel, whose work forms the basis for Chapter 5, their differential-space theory of quantum systems carries the Bohr-Einstein controversy "to a new stage of concreteness."

It realizes the formal program of Einstein: namely, a physical picture in which a collection of identically prepared (subject to the indeterminacy principle) systems is represented by an ensemble of different, precisely defined systems so that the dispersion in values of measured quantities has its counterpart in the variety of properties of the systems in the ensemble.

The differential-space theory of quantum systems is now characterized as it will be given in Chapter 5.

It is a probability theory using as its sample space the differential space associated with the Hilbert space of the quantum-mechanical description of a system. It is postulated that the realized values of an observable must be eigenvalues of its associated hypermaximal operator. There is an algorithm (actually a choice of alternative algorithms) for the assignment of realized values of an observable to points in differential space, which amounts to formulating observables as functions of the (Brownian motion) function $X(t, \alpha)$; it can then be proved that the distribution function of any observable as so constructed agrees with that of quantum mechanics.

In contrast to the treatment of the wave function in the Wiener-Siegel theory as representing an *ensemble* of individuals, it is important to examine the more familiar Bohr point of view, in which the wave function represents an individual system. It will be emphasized in Chapter 5 that the ensemble approach enables one, in large part, to use the logic of measure-theoretic probability in quantum-theoretic analysis, though it is not possible to clarify at this time what new experimental interpretation the measure-theoretic logic might hold for such experiments as the two-slit electron diffraction experiment. Because of the divergence in spirit between the Bohr treatment of the individual quantum mechanical system and the conception of measure-theoretic probability, we may expect the future contributions of hidden variable theories to shed light on the universal tenability of completely additive probability measures. A brief indication of the relevance of probability functionals (that are not necessarily completely additive) to the treatment of individual systems was given at the end of Section 2.

We quote Feynman's [1951] discussion of the fundamentally unpredictable nature of certain experimental outcomes:

But far more fundamental was the discovery that in nature the laws of combining probabilities were *not* those of the classical probability theory of Laplace. . . . We shall see that the quantum mechanical laws of the physical world approach very closely the laws of Laplace as the size of the objects involved in the experiments increases. Therefore, the laws of probabilities which are conventionally applied are quite satisfactory in analyzing the behavior of the roulette wheel but not the behavior of a *single electron or a photon of light.*

In reference to the two-slit electron diffraction experiment, Feyman continues:

We *compute* the intensity (that is the absolute square of the amplitude) of waves which would arrive in the apparatus at X, and then *interpret* this intensity as the probability that a particle will arrive at X. . . . To discuss this point in more detail we first consider the situation which arises from the observation that our new law . . . of composition of probabilities implies in general, that it is *not true* that $P = P_1 + P_2$. . . . When no attempt is made to determine through which hole the electron passes, one cannot say it must pass through one hole or the other. Only in a situation where an apparatus is operating to determine which hole the electron goes through is it permissible to say that it passes through one or the other. . . .

To summarize then: The probability of an event (in an ideal experiment where there are no uncertain external disturbances) is the absolute square of a complex quantity called the probability amplitude. When the event can occur in several alternative ways the probability amplitude is the sum of the probability amplitude for each alternative considered separately.

If an experiment capable of determining which alternative is actually taken is performed, the interference is lost and the probability becomes the sum of the probability for each alternative.

In Chapter 5, it is pointed out, in reference to the two-slit experiment, that

In order to realize a description of quantum systems in which all observables are defined, one must do "something more" than assign (in imagination) sharp values to two conjugate variables of a system. The classic way to do this "something more" is to add further determining —"hidden"—variables to the system; these can then serve the all-important purpose of expressing the influence of "this slit" even when the particle is passing through "the other slit."

Thus we verify that the hidden variable theories of quantum mechanics bear not only upon questions of physical concern, such as the electron diffraction phenomenon, but also upon mathematical questions, such as the appropriateness of the complete additivity postulate in a universally applicable probability theory.

Among the statistical postulates that are used in the differential-space theory of quantum systems is the condition that "the measure of a suitably defined subensemble is never negative." Though Wiener and Siegel correctly refer to this condition in distinguishing the results of their work from the phase-space results of Wigner [1932], Moyal [1949], and others, certain recent work (Rankin [1966b]) has added the following information. At least in the case of the harmonic oscillator, the Born statistical postulate (identified with the assumption that the function F_R in Equation 32 is a probability distribution function for the observable R) may be stronger than necessary to bring agreement between experimental results and ordinary quantum-theoretic calculations. In fact, a slight weakening of the Born postulate has made it possible without the use of differential space to deduce quantum-mechanical phase-space distributions that do, indeed, satisfy the stated condition that measures on subensembles should never be negative. These remarks simply serve to illustrate the richness of unexplored approaches to the kinds of questions opened up by the differential-space approach to quantum theory. It is clearly stated in Chapter 5 that "Any ultimate physically interesting formulation of hidden variables may well differ considerably from this one (the Wiener-Siegel formulation). If this is true, it may be that the latter is to be regarded mainly as a means of proving, by construction, the possibility of a theory of hidden variables consistent with quantum phenomena."

2

The Brownian Motion Process
and Differential Space

I. The Brownian motion

In 1827 the English botanist Robert Brown first observed under the microscope the tiny irregular motions of small particles suspended in water. This motion is known today as *Brownian motion*. Microscopic particles immersed in a fluid move under the impacts of the molecules of the fluid in such a way that the future behavior of the particle is independent of its past behavior. It is not, however, the position of a particle which is independent of its past but rather the *change* of position of the particle between two times.

Suppose we take the path of a particle subject to Brownian motion and consider one of its coordinates, say the x coordinate, as a function of the time t, $x = X(t)$. The fact that it is the amount by which $X(t)$ changes in a time interval $t_1 < t < t_1 + h$, which is independent of the value of $X(t)$ for $t \le t_1$, inspired the comment by Paley and Wiener [1934, p. 141]: "In crude language, [it is] the differentials of the function $X(t)$ [that] correspond to the independent coordinates of a point in space of a finite number of dimensions, and not the values of the function."

Various investigations of the physical and mathematical properties of Brownian motion have been made. The main contributions prior to 1923 were by Einstein [1905; 1906], Perrin [1910], and Smoluchowski [1916; 1918]. Bachelier's contributions [1900], while viewed today as mathematically deficient, still have bearing on the subject. The stochastic process, which we will call the *Brownian*

33

motion process, was first given a complete mathematical formulation by Wiener [1923] in his paper on differential space. This process is often called the Wiener process.

Before describing the Brownian motion process mathematically, we will make a few further remarks about the (physical) motion of a Brownian particle and simplifying approximations that lead to the definition of this process. We are considering the motion of a *free particle*, that is, one in which there is no external field of force. Strictly, in order to know the motion of the particle, one would have to know not only the impulses which it receives over a given time but the initial velocity of the particle. In the Einstein theory of this motion, this initial velocity over any interval of time is of negligible importance in comparison with the impulses received during the same interval of time. (It is known today that under normal conditions a Brownian particle will suffer about 10^{21} collisions per second.) Under the customary reasonable physical assumption of a random distribution of molecular positions and velocities, and under the further assumption that the intervals of time under consideration are large with respect to the intervals between molecular collisions, it follows,[1] as Einstein has pointed out, that the probability that $X(t_1) - X(t)$, $t_1 > t$, lie between a and b is very nearly of the form

$$\frac{1}{\sqrt{\pi A(t_1 - t)}} \int_a^b \exp\left[-\frac{\xi^2}{A(t_1 - t)} \right] d\xi, \tag{1}$$

where

$$A = \frac{4RT}{Nf}, \tag{2}$$

with R the universal gas constant, T the absolute temperature, N Avogadro's number, and f the friction coefficient. (In the case of a spherical particle to which Stokes's law applies, f can be expressed in terms of the size of the particle and the viscosity of the fluid containing the suspension, yielding $A = \frac{2}{3}(KT/\pi a \eta)$, where K is Boltzmann's constant R/N, a is the radius of the sphere, and η is the coefficient of viscosity of the surrounding fluid.)

---

[1] A more exact physical treatment (see Uhlenbeck and Ornstein [1930]), leads to a more complicated expression, of which that given here is the asymptotic limit in time; the present form would also be obtained in the limit as $m/f \to 0$, where m is the mass of the particle and f is the coefficient of friction.

As Kac [1947, p. 370] points out,

The greatness of Einstein's contribution was not, however, solely due to the derivation of [the diffusion equation

$$\frac{\partial P}{\partial t} = \frac{A}{4} \frac{\partial^2 P}{\partial x^2} \tag{3}$$

for the probability density $P(x; t)$ which leads to 1]. From the point of view of physical applications, it was equally, or perhaps even more important that he was able to [obtain the relation 2]. It was relation 2 that made possible the determination of Avogadro's number from Brownian motion experiments, an achievement for which Perrin was awarded the Nobel prize in 1926.

We venture to suggest that even this does not do full justice to Einstein's achievement. It was through this work that statistical mechanics (at the time called kinetic theory) first emerged as a theory having experimental consequences independent of those of thermodynamics. The validity of the experimental predictions necessarily gave powerful support to the atomic hypothesis on which they were based. At the present time it is hard to appreciate the strength of the opposition to the atomic hypothesis that existed at that time (cf. Mach [1923, p. 364]). Einstein's theory of fluctuations, inaugurated by this work, has yielded fruits that are only beginning to appear in physical theory; it is one of the cornerstones of the theory of noise in electric circuits (and elsewhere), and of the thermodynamics of stationary irreversible processes. We may quote from the first of Einstein's papers on the Brownian movement:

If the movement discussed here can actually be observed . . ., then classical thermodynamics can no longer be looked upon as applicable with precision to bodies even of dimensions distinguishable in a microscope: an exact determination of actual atomic dimensions is then possible. On the other hand, had the prediction of this movement proved to be incorrect, a weighty argument would be provided against the molecular-kinetic conception of heat.

2. The Brownian motion process

The work of Einstein, Smoluchowski, and Wiener leads one to approximate the x coordinate $X(t)$ of (physical) Brownian motion particles with the sample functions of a Brownian motion process. In this and the next sections, we shall define the mathematical function of Brownian motion and shall study its properties. Today there are many ways to do this. Essentially, we will follow Wiener's initial method in which he constructed an explicit mapping of the space C of real-valued functions which are continuous in $0 \leq t \leq 1$ and vanish at $t = 0$, into the unit interval (except for sets of measure zero). Treatments more in the spirit of abstract measure theory have been given, for example, by Doob and Lévy. A comparison of Doob's method and that of Wiener can be found in Kac [1959, pp. 161–164].

In this consideration we first need the formula for the composition of Gaussian distributions that are independent over independent intervals of time. For this purpose, consider a quantity X that has a Gaussian distribution. The probability that this quantity lies between x_1 and x_2 is given by

$$P = \int_{x_1}^{x_2} \frac{1}{\sqrt{2\pi a}} \, e^{-x^2/2a} \, dx. \tag{4}$$

For the Brownian motion functions we want the quantity a to depend upon the time variable t in a suitable way. The composition formula will suggest this way. If we think of expression 4 as referring to the position at a given time, and if we consider the displacements of the particles at a second time, starting from those positions, then the probability density is given by

$$\frac{1}{\sqrt{2\pi a}} \, e^{-x^2/2a} \, \frac{1}{\sqrt{2\pi b}} \, e^{-(y-x)^2/2b}, \tag{5}$$

where the term $y - x$ comes in because we are measuring displacements. The probability density for y, when x ranges over all possible values, is given by

$$\int_{-\infty}^{\infty} \frac{1}{\sqrt{2\pi a}} \, e^{-x^2/2a} \, \frac{1}{\sqrt{2\pi b}} \, e^{-(y-x)^2/2b} \, dx. \tag{6}$$

An easy calculation shows that Equation 6 is equal to

$$\frac{1}{\sqrt{2\pi(a+b)}} \, e^{-y^2/2(a+b)}, \tag{7}$$

and this gives the desired law of composition of Gaussian distributions. We note that the parameter a adds when we compose two Gaussian distributions.

If we make use of the property that the motion is independent over nonoverlapping time intervals, and that the displacement is dependent only on the length of the time interval and not on the original time, then the parameter a in expression 4 should be a linear function of the time difference t. We normalize it so that it is t itself. Thus we are led to choose the density function

$$\frac{1}{\sqrt{2\pi t}} \, e^{-x^2/2t}. \tag{8}$$

Before we can investigate various properties of the functions of the mathematical Brownian motion, we must first define the functions with which we shall be dealing, and we want to establish an integration theory. To do this we first set up a mapping between certain sets of functions called quasi-intervals, and certain subintervals of the unit interval $0 \leq \alpha \leq 1$. We do this in such a manner that we will have a class of functions $X(t, \alpha)$ on $0 \leq t \leq 1, 0 \leq \alpha \leq 1$, which are measurable in (t, α) and which for almost every α are continuous in t. These will be the functions of our Brownian motion.

We now consider the class of all real-valued functions $x(t)$ on the unit interval $0 \leq t \leq 1$ and vanishing at $t = 0$. A quasi-interval is defined as follows: Let n be any positive integer, and let $t_1, \cdots, t_n$ be n points on the unit interval of the time axis, with

$$0 < t_1 < t_2 < \cdots < t_n \leq 1.$$

Through each of these n points we pass a line perpendicular to the t-axis, and on each such line we choose an interval (λ_j, μ_j), for $j = 1, \cdots, n$, with

$$-\infty \leq \lambda_j < \mu_j \leq \infty.$$

A quasi-interval $I(n; t_1, \cdots, t_n; \lambda_1, \mu_1, \cdots, \lambda_n, \mu_n)$ consists of all real-valued functions $x(t)$ on $0 \leq t \leq 1$ with $x(0) = 0$, such that

$$\lambda_j < x(t_j) \leq \mu_j; \qquad j = 1, \cdots, n. \tag{9}$$

The work of the preceding section leads us to associate a number with the quasi-interval 9, which we call the probability associated with $I(n; t_1, \cdots, t_n; \lambda_1, \mu_1, \cdots, \lambda_n, \mu_n)$, namely

$$P\{I\} = \left[\frac{1}{\sqrt{(2\pi)^n t_1(t_2 - t_1) \cdots (t_n - t_{n-1})}} \right]$$
$$\times \int_{\lambda_1}^{\mu_1} \cdots \int_{\lambda_n}^{\mu_n} \exp\left[-\sum_{j=1}^{n} \frac{(u_j - u_{j-1})^2}{2(t_j - t_{j-1})} \right] du_1 \cdots du_n, \quad (10)$$

where for convenience we have used the notation $t_0 = u_0 = 0$.

Clearly, if the class of all such functions $x(t)$ be divided into a finite number of quasi-intervals, some of which must then contain infinite values of λ_j or μ_j, the sum of their probabilities will be one. Also, if we insert a value t^* between two consecutive values in the subdivision, say $t_j < t^* < t_{j+1}$, and if the associated interval (λ^*, μ^*) is chosen to be $(-\infty, \infty)$, then the probability of the new quasi-interval

$$I(n + 1; t_1, \cdots, t_j, t^*, t_{j+1}, \cdots, t_n;$$
$$\lambda_1, \mu_1, \cdots, \lambda_j, \mu_j, -\infty, \infty, \lambda_{j+1}, \mu_{j+1}, \cdots, \lambda_n, \mu_n)$$

will be the same as that of the original quasi-interval 9.

In our mapping we shall consider not all such quasi-intervals, but only certain ones built up from the binary points of the unit interval $0 < t \leq 1$. The ones with which we will work are defined as follows: For each positive integer n, and for each (2^n)-tuple

$$(m_1, \cdots, m_{2^n}) \qquad (11)$$

with

$$m_j = -2^{n-1}, -2^{n-1} + 1, \cdots, 2^{n-1} - 1; \qquad j = 1, \cdots, 2^n \quad (12)$$

we define a quasi-interval

$$Q(n; m_1, \cdots, m_{2^n}): \qquad \tan \frac{m_j \pi}{2^n} < x\left(\frac{j}{2^n}\right) \leq \tan \frac{(m_j + 1)\pi}{2^n};$$
$$j = 1, \cdots, 2^n. \quad (13)$$

For each n we have $(2^n)^{2^n}$ of these quasi-intervals. Also, each quasi-interval 13 is made up of a finite number of quasi-intervals $Q(n + 1; l_1, \cdots, l_{2^{n+1}})$, and the sum of the probabilities belonging to the latter group gives the probability belonging to the former. It is with this set of quasi-intervals that we do our mapping.

We first consider the four quasi-intervals for $n = 1$, namely

$$Q(1; -1, -1), \qquad Q(1; -1, 0), \qquad Q(1; 0, -1), \qquad Q(1; 0, 0),$$

and we map these on the unit interval as follows: Starting at the origin O we lay out on the α-axis an interval OB whose length is $P\{Q(1; -1, -1)\}$, and as one part of our mapping, we say that the segment OB of the α-axis corresponds to the quasi-interval $Q(1; -1, -1)$. We next lay off on the α-axis an interval BC (with C to the right of B) whose length is $P\{Q(1; -1, 0)\}$, and similarly for the other two quasi-intervals $Q(1; 0, -1)$, $Q(1; 0, 0)$. This gives us four segments OA, AB, BC, CD of the interval $[0, 1]$, that is, four probabilities, which add up to unity. This is the first stage of our mapping.

We next map the $4^4 = 256$ quasi-intervals $Q(2; m_1, \cdots, m_4)$ into 256 intervals of the segment $0 \le \alpha \le 1$, translating probability into length, and in such a manner that if $Q(2; l_1, \cdots, l_4)$ forms a portion of $Q(1; m_1, m_2)$, their intervals stand in the same relation. If we keep up this process indefinitely, we shall have mapped all the quasi-intervals $Q(n; m_1, \cdots, m_{2^n})$ into intervals of $0 \le \alpha \le 1$ in such a way that probability is translated into length, and the interval on $0 \le \alpha \le 1$ corresponding to any one quasi-interval $Q(n; m_1, \cdots, m_{2^n})$ will be composed of subintervals corresponding precisely to those quasi-intervals $Q(n + 1; l_1, \cdots, l_{2^{n+1}})$ which make up $Q(n; m_1, \cdots, m_{2^n})$.

It is easily seen that the lengths of the image intervals on $0 \le \alpha \le 1$ will go to zero as n approaches infinity.

If we consider any value of α in $0 \le \alpha \le 1$ which is not an end point of one of this denumerable set of subdivisions of the interval $[0, 1]$, this value of α will lie in one of the four intervals which we obtained at the first stage of our mapping, in one of the 256 intervals which we obtained at the second stage, and so on. Except for these boundaries which form a set of Lebesgue measure zero, it will be uniquely determined by where it is at the different stages of subdivision. Thus every point of $0 \le \alpha \le 1$, except for a set of measure zero, is uniquely characterized by and uniquely characterizes an infinite sequence of intervals, each containing the next, and tending to zero in length. Also, by our mapping, every point of $0 \le \alpha \le 1$, except for a set of measure zero, is uniquely characterized by and characterizes an infinite sequence of quasi-intervals of the class 13, each containing the next, and having probabilities which tend to zero.

Now consider a point of the interval $0 \leq \alpha \leq 1$ which is not an end point of one of the subintervals into which our mapping divided the segment. We might ask whether the sequence of quasi-intervals associated with this point defines a unique function. We can certainly say the following: If $x_1(t)$ and $x_2(t)$ are two functions common to the sequence of quasi-intervals in question, then $x_1(t) = x_2(t)$ for all binary points t in $0 \leq t \leq 1$. Thus we can conclude that if there is a continuous function it is unique.

In the next section, we shall prove the following result.

Theorem I   *Let λ be fixed in $0 \leq \lambda < \frac{1}{2}$ and let h be any positive number. Let R be the set of real-valued functions $x(t)$ on $0 \leq t \leq 1$ which vanish at the origin. The subset of functions of R which satisfy the inequality*

$$|x(t_2) - x(t_1)| > h|t_2 - t_1|^\lambda \tag{14}$$

for at least one pair of binary numbers in $0 \leq t \leq 1$ can be enclosed in a (denumerable) number of quasi-intervals of the collection $Q(v; m_1, \cdots, m_{2^v})$ the sum of whose probabilities is less than $l(h)$, where

$$\lim_{h \to \infty} l(h) = 0. \tag{15}$$

For the present we will assume this result. If $\epsilon > 0$, then there is a real number $h_0 = h_0(\epsilon)$ such that $l(h) < \epsilon$ for $h > h_0$. If we then delete a denumerable set of quasi-intervals of total probability less than ϵ, the remaining quasi-intervals and portions of quasi-intervals all contain functions of R satisfying the condition

$$|x(t_1) - x(t_2)| \leq h|t_1 - t_2|^\lambda \tag{16}$$

for all binary points t_1, t_2 in $0 \leq t \leq 1$. To each such function $x(t)$ we may associate a unique continuous function $X(t)$, namely the function defined by $X(t) = x(t)$ when t is a binary point, and $X(t) = \lim_{t_n \to t} x(t_n)$ otherwise, where t_n is a sequence of binary points with $\lim t_n = t$. This associated continuous function satisfies condition 16 for every pair of points t_1, t_2 in $0 \leq t \leq 1$. We can now use

Ascoli's theorem. The functions $X(t)$ satisfying condition 16 form an equicontinuous class, and they are uniformly bounded (by h, as one sees on choosing $t_1 = 0$). Thus, any sequence of functions $X(t)$ has a uniformly convergent subsequence. The limit of this subsequence is itself a continuous function, and indeed it satisfies the same condition of equicontinuity, namely expression 16, for all pairs t_1, t_2.

Thus if we modify the interval $0 \le \alpha \le 1$ by the removal of a set of points of arbitrarily small outer measure, as well as by the removal of the end points of our intervals, every point α which remains is characterized by a sequence of intervals closing down on it, by the succession of corresponding quasi-intervals, and by the uniquely determined function $X(t)$ common to this sequence of quasi-intervals and satisfying condition 16 for all values of t_1, t_2. We denote this function by $X(t, \alpha)$. This is our stochastic function which defines the Brownian motion process. By studying the mapping and making use of the equicontinuity condition 16 one sees that $X(t, \alpha)$ is a measurable[2] function of (t, α). It satisfies a condition of the form given by 16 for all values of t_1 and t_2, for almost all α. Thus, in particular, for almost all α it is a continuous function of t. Any functional of the $X(t, \alpha)$ determines a function on the line $0 \le \alpha \le 1$, which may be Lebesgue summable. In the latter case, we shall define the average or expected value of the functional as the Lebesgue integral of the corresponding function on $0 \le \alpha \le 1$.

Among the summable functionals are the expressions

$$G[X(t_1, \alpha), \cdots, X(t_n, \alpha)], \tag{17}$$

where G is a polynomial. If

$$0 < t_1 < t_2 < \cdots < t_n \le 1, \tag{18}$$

---

[2] The measurability in (t, α) of $X(t, \alpha)$ can also be seen as a consequence of the following theorem [see Doob, 1953, p. 60]. "Let $\{x_t(\omega), t \in T\}$ be a separable process with a Lebesgue measurable parameter set T. Suppose that there is a t set T_1 of Lebesgue measure zero such that

$$P\left\{ \lim_{s \to t} x_s(\omega) = x_t(\omega) \right\} = 1, \qquad t \in T - T_1.$$

Then the x_t process is measurable;" i.e., $x_t(\omega)$ defines a function measurable in the pair of variables (t, ω).

then the average, or expected value, of expression 17, which we may write as

$$E\{G[X(t_1, \alpha), \cdots, X(t_n, \alpha)]\} \tag{19a}$$

or as

$$\int_0^1 G[X(t_1, \alpha), \cdots, X(t_n, \alpha)] \, d\alpha, \tag{19b}$$

is given by

$$\left[\sqrt{(2\pi)^n t_1(t_2 - t_1) \cdots (t_n - t_{n-1})}\right]^{-1} \int_{-\infty}^{\infty} \cdots \int_{-\infty}^{\infty} G(u_1, \cdots, u_n)$$

$$\times \exp\left[-\sum_{j=1}^{n} \frac{(u_j - u_{j-1})^2}{2(t_j - t_{j-1})}\right] du_1 \cdots du_n. \tag{19c}$$

For convenience we have used the notation

$$t_0 = u_0 = 0. \tag{20}$$

The summability of expression 17 and the equality of 19c with expressions 19a and b is readily seen if the t_j are binaries. The extension to other values of the t_j follows from the equicontinuity property, condition 16. More generally, if the integral in expression 19c is Lebesgue summable, then the functional 17 is summable and expressions 19a and b are equal to 19c. If we denote by C the set of all functions $X(t, \alpha)$ of our process, and if $\Phi[X(\cdot, \alpha)|t]$ is any summable functional over C, then we write

$$E\{\Phi[X(\cdot, \alpha)|t]\} = \int_0^1 \Phi[X(\cdot, \alpha)|t] \, d\alpha.$$

If A is any measurable subset of C and if $\Phi[X(\cdot, \alpha)|t]$ is summable over A, then the probability that $X(t, \alpha)$ be such that $\Phi[X(\cdot, \alpha)|t]$ lies in A is of course merely the measure of the set of values of α for which $\Phi[X(\cdot, \alpha)|t]$ lies in A.

By our mapping, we see that if S is a Lebesgue measurable set of points in Euclidean n-space, if 18 holds, and if A_S is the set of functions $X(t, \alpha)$ for which

$$[X(t_1, \alpha), \cdots, X(t_n, \alpha)] \in S \tag{21}$$

then the set A_S is measurable and

$$m(A_S) = \int_{A_S} d\alpha = P\{[X(t_1, \alpha), \cdots, X(t_n, \alpha)] \in S\}$$

$$= \frac{1}{\sqrt{(2\pi)^n t_1(t_2 - t_1) \cdots (t_n - t_{n-1})}}$$

$$\times \int \cdots \int_S \exp\left[-\sum_{j=1}^{n} \frac{(u_j - u_{j-1})^2}{2(t_j - t_{j-1})}\right] du_1 \cdots du_n, \qquad (22)$$

where we have again used the convention 20. For each t in $0 < t \leq 1$ and each real number γ we have

$$P\{X(t, \alpha) < \gamma\} = \frac{1}{\sqrt{2\pi t}} \int_{-\infty}^{\gamma} e^{-u^2/2t} \, du, \qquad (23)$$

and thus for each t, the variable $X(t, \alpha)$ is Gaussian with mean zero and variance t. We shall show that for each pair of distinct values t_1, t_2 in $[0, 1]$ the increment $X(t_2, \alpha) - X(t_1, \alpha)$ is a Gaussian variable with mean zero and variance $|t_2 - t_1|$, and we shall further show that the Brownian motion process has independent increments. To show these two properties, let $t_1, \cdots, t_n$ be any points satisfying condition 18 and let $\gamma_1, \cdots, \gamma_{n-1}$ be any real numbers. For each index $j = 1, \cdots, n - 1$, the set of points α on $0 \leq \alpha \leq 1$ for which

$$X(t_{j+1}, \alpha) - X(t_j, \alpha) < \gamma_j \qquad (24)$$

is measurable with measure

$$P\{X(t_{j+1}, \alpha) - X(t_j, \alpha) < \gamma_j\}$$

$$= [\sqrt{(2\pi)^2 t_j(t_{j+1} - t_j)}]^{-1} \int_{-\infty}^{\infty} \int_{-\infty}^{u_j + \gamma_j} \exp\left[-\frac{u_j^2}{2t_j} - \frac{(u_{j+1} - u_j)^2}{2(t_{j+1} - t_j)}\right]$$

$$\times \, du_j \, du_{j+1}. \qquad (25)$$

Now the integral

$$\int_{-\infty}^{u_j + \gamma_j} \exp\left[-\frac{(u_{j+1} - u_j)^2}{2(t_{j+1} - t_j)}\right] du_{j+1} \qquad (26)$$

is independent of u_j, and hence we have

$$P\{X(t_{j+1}, \alpha) - X(t_j, \alpha) < \gamma_j\}$$

$$= \frac{1}{\sqrt{2\pi(t_{j+1} - t_j)}} \int_{-\infty}^{u_j + \gamma_j} \exp\left[-\frac{(u_{j+1} - u_j)^2}{2(t_{j+1} - t_j)}\right] du_{j+1}, \qquad (27)$$

whatever (real) value u_j has. This yields the first property, namely that the increment $X(t_{j+1}, \alpha) - X(t_j, \alpha)$ is a Gaussian variable with mean zero and variance $|t_{j+1} - t_j|$. It also enables us to show that the process has independent increments.

To prove the latter we need to show that

$$P\{X(t_2, \alpha) - X(t_1, \alpha) < \gamma_1, \cdots, X(t_n, \alpha) - X(t_{n+1}, \alpha) < \gamma_{n-1}\}$$
$$= P\{X(t_2, \alpha) - X(t_1, \alpha) < \gamma_1\} \cdots P\{X(t_n, \alpha) - X(t_{n-1}, \alpha) < \gamma_{n-1}\}. \quad (28)$$

This will show that the differences $X(t_2, \alpha) - X(t_1, \alpha), \cdots, X(t_n, \alpha) - X(t_{n-1}, \alpha)$ are mutually independent increments. The left-hand side of Equation 28 is given by

$$\frac{1}{\sqrt{(2\pi)^n t_1(t_2 - t_1) \cdots (t_n - t_{n-1})}}$$
$$\times \int_{-\infty}^{\infty} \int_{-\infty}^{u_1 + \gamma_1} \cdots \int_{-\infty}^{u_{n-1} + \gamma_{n-1}} \exp\left[-\sum_{j=0}^{n-1} \frac{(u_{j+1} - u_j)^2}{2(t_{j+1} - t_j)} \right] du_1 \cdots du_n, \quad (29)$$

and by Equation 27 this is equal to the product of

$$(2\pi t_1)^{-\frac{1}{2}} \int_{-\infty}^{\infty} e^{-u^2/2t_1} \, du_1 \quad (30)$$

times the right-hand side of Equation 28. Since expression 30 has the value one, this yields Equation 28.

In the next section we shall prove Theorem 1, concerning equicontinuity, which we have been using in this section. We shall then show that almost no sample function of the Brownian motion process has bounded variation. Even though this is the case, Wiener has shown how to give an interpretation to the expression

$$\int_0^1 f(t) \, dX(t, \alpha)$$

for almost all α, where $f(t) \in L_2(0, 1)$. We shall present this work in Section 1 of Chapter 3.

It might be well to mention at this stage that the actual definition of the Brownian motion process is sometimes given directly in probability terms as follows: "It is a real process $\{X(t), t \in T\}$ where

T is usually taken to be an interval, in fact, usually either $(-\infty, \infty)$ or $[0, \infty)$, with independent Gaussian increments satisfying

$$E\{X(t) - X(s)\} = 0, \quad E\{[X(t) - X(s)]^2\} = \sigma^2 |t - s|$$

where σ is a positive constant." (See Doob [1953, pp. 97–98, 392–398].) One proves directly from this that almost all sample functions of a separable Brownian motion process are continuous, and that almost no sample function has bounded variation.

While it is easy to map the Brownian motion process defined on $0 \leq t \leq 1$ to one defined on $[0, \infty)$ or on $(-\infty, \infty)$ (see, for example, Wiener [1930]), we shall not give the details of the mapping here. Rather, when we wish to consider the Brownian motion process on the infinite or semi-infinite interval, or more generally for $t \in T$, we shall understand that we have a separable real process $\{X(t), t \in T\}$ with independent Gaussian increments $X(t) - X(s)$ that have mean zero and variance $|t - s|$. We emphasize that the Brownian motion process defined by an original mapping is separable, and we consider only separable Brownian motion processes.

3. The equicontinuity property

We shall prove the equicontinuity Theorem 1, which was used in the preceding section. We shall give the proof in stages.

Let m be a positive integer and consider the 2^m intervals

$$(-a_p, -a_{p-1}), (-a_{p-1}, -a_{p-2}), \cdots, (-a_1, 0), (0, a_1), \cdots, (a_{p-1}, a_p), \tag{31}$$

where

$$p = 2^{m-1}; \quad a_i = \tan \frac{\pi i}{2^m}; \quad i = 0, 1, \cdots, p. \tag{32}$$

(We do not denote explicitly the dependence of p and the a_i upon m.)

If E and ϵ are any positive numbers, then clearly there exists a positive integer $m_0 = m_0(E, \epsilon)$ such that for $m > m_0$ each of the intervals in the collection 31 which has at least one end point in $(-E, E)$ has length less than ϵ. This means that if $a_s < E$ for some index s, then $a_{s+1} - a_s < \epsilon$. We denote by R_0 the class of all real-valued functions $x(t)$ defined in $0 \leq t \leq 1$ which vanish at $t = 0$.

We shall now prove the following lemma:

Lemma I *Let n be a positive integer, let j be a nonnegative integer $< 2^n$, let $0 < \lambda < \frac{1}{2}$, and let $h > 2$. The set of functions $x(t)$ of R_0 for which*

$$\left| x\left(\frac{j+1}{2^n}\right) - x\left(\frac{j}{2^n}\right) \right| > h2^{-n\lambda} \tag{33}$$

can be enclosed in a finite set of quasi-intervals of the collection $Q\{\nu; m_1, \cdots, m_{2^\nu}\}$, the sum of whose probabilities is less than

$$\frac{1}{h2^{2n}} + 2 \exp\left[-\frac{h}{2} 2^n(\frac{1}{2} - \lambda) \right]. \tag{34}$$

Proof. First let E be a positive number such that

$$\frac{4}{\sqrt{2\pi}} \int_E^\infty e^{-v^2/2}\, dv < \frac{1}{h2^{2n}}, \tag{35}$$

and let ϵ be a positive number satisfying the inequality

$$\epsilon < \frac{h}{8} 2^{-n\lambda}. \tag{36}$$

With these values of E and ϵ, consider the 2^m intervals in the collection 31 with $m > m_0(E, \epsilon)$. We also assume that m is greater than the positive integer n of the lemma. We divide this set of intervals into two subsets, namely,

(*a*) those which have at least one end point in $(-E, E)$, and
(*b*) those completely outside $(-E, E)$.

If s is the largest integer such that $a_{s-1} < E$, then the intervals in the subset (*a*) are the $(2s)$ intervals

$$(-a_s, -a_{s-1}), \cdots, (-a_1, 0), (0, a_1), \cdots, (a_{s-1}, a_s), \tag{37}$$

which we shall relabel as

$$(b_\sigma, b_{\sigma+1}); \qquad \sigma = 1, \cdots, 2s. \tag{38}$$

We now proceed with the proof of the lemma. For convenience in notation, we write

$$t_1 = \frac{j}{2^n}, \quad t_2 = \frac{j+1}{2^n}, \quad A = h(t_2 - t_1)^\lambda = h2^{-n\lambda}$$

$$f(u_1, u_2) = \exp\left[-\frac{u_1^2}{2t_1} - \frac{(u_2 - u_1)^2}{2t_2 - 2t_1}\right]. \tag{39}$$

We will prove the lemma only for $j > 0$. The proof for $j = 0$ is much simpler, and will be evident to the reader who follows the proof for $j > 0$.

The quasi-intervals that we will use to enclose the functions of R_0 satisfying condition 33 will be of three types, namely,

   i. the $2(p - s)$ quasi-intervals for which $x(t_1)$ lies in one of the $2(p - s)$ intervals of class (b),
   ii. the $2(p - s)$ quasi-intervals for which $x(t_2)$ lies in one of the $2(p - s)$ intervals of class (b),
   iii. a selected number of the $(2s)^{2s}$ quasi-intervals, namely,

$$I_{\mu, \nu} : b_\mu < x(t_1) \le b_{\mu+1}, \, b_\nu < x(t_2) \le b_{\nu+1},$$

where μ and ν range independently over $1, \cdots, 2s$.

We note that the quasi-intervals considered in i, ii, and iii are all quasi-intervals of the class $Q(\nu; m_1, \cdots, m_{2^\nu})$ defined in statement 13, with $\nu = m$, where m is a fixed integer greater than max $\{m_0(E, \epsilon), n\}$.

The sum of the probabilities of the $4 (p - s)$ quasi-intervals in (i) and (ii) is less than

$$\frac{2}{\sqrt{2\pi t_1}} \int_E^\infty e^{-u^2/2t_1} \, du + \frac{2}{\sqrt{2\pi t_2}} \int_E^\infty e^{-u^2/2t_2} \, du$$

$$= \frac{2}{\sqrt{2\pi}} \int_{Et_1^{-1/2}}^\infty e^{-v^2/2} \, dv + \frac{2}{\sqrt{2\pi}} \int_{Et_2^{-1/2}}^\infty e^{-v^2/2} \, dv$$

$$\le \frac{4}{\sqrt{2\pi}} \int_E^\infty e^{-v^2/2} \, dv < \frac{1}{h2^{2n}}, \tag{40}$$

where, in the final step, we have used expression 35.

Before considering the quasi-intervals $I_{\mu, \nu}$ of iii, let us make the following general observation. If $c < x(t_1) \le d$, then the inequality

$x(t_2) - x(t_1) > A$ implies $x(t_2) > A + c$, and the inequality $x(t_2) - x(t_1) < -A$ implies $x(t_2) < -A + d$. Since all the intervals in the collection 38, which are the ones used in defining the $I_{\mu, \nu}$, are of length less than ϵ, we see that if

$$b_\mu < x(t_1) \le b_{\mu+1}, \tag{41}$$

then all the intervals of the collection 38 which contain $x(t_2)$ must lie entirely in the union of the two intervals

$$(-\infty, -A + b_{\mu+1} + \epsilon), \quad (A + b_\mu - \epsilon, \infty), \tag{42}$$

and hence the sum of the probabilities of all the quasi-intervals $I_{\mu, \nu}$ which contain functions satisfying condition 33 is less than

$$\frac{1}{\sqrt{(2\pi)^2 t_1 (t_2 - t_1)}} \sum_{\mu=1}^{2s} \int_{b_\mu}^{b_{\mu+1}} \left[\int_{-\infty}^{-A + b_{\mu+1} + \epsilon} + \int_{A + b_\mu - \epsilon}^{\infty} \right] f(u_1, u_2) \, du_1 \, du_2$$

$$\le \frac{1}{\sqrt{(2\pi)^2 t_1 (t_2 - t_1)}} \sum_{\mu=1}^{2s} \int_{b_\mu}^{b_{\mu+1}} \left[\int_{-\infty}^{-A + u_1 + 2\epsilon} + \int_{A + u_1 - 2\epsilon}^{\infty} \right] f(u_1, u_2) \, du_1 \, du_2$$

$$\le \frac{2}{\sqrt{(2\pi)^2 t_1 (t_2 - t_1)}} \int_{-E-\epsilon}^{E+\epsilon} e^{-u_1^2/2t_1} \, du_1 \int_{A-2\epsilon}^{\infty} e^{-v^2/2(t_2 - t_1)} \, dv$$

$$\le \frac{2}{\sqrt{2\pi(t_2 - t_1)}} \int_{A-2\epsilon}^{\infty} e^{-[v^2/2(t_2 - t_1)]} \, dv = \frac{2}{\sqrt{\pi}} \int_{(A-2\epsilon)/\sqrt{2(t_2 - t_1)}}^{\infty} e^{-x^2} \, dx$$

$$\le 2 \int_{A/2\sqrt{t_2 - t_1}}^{\infty} e^{-x} \, dx = 2e^{-A/2\sqrt{t_2 - t_1}} = 2e^{-(h/2)2^n(\frac{1}{2} - \lambda)} \tag{43}$$

In passing from the next-to-last integral to the last integral we have used the inequality 36 and the inequalities $\sqrt{2} < \frac{3}{2}$ and $x < x^2$ for $x > 1$.

Adding the final terms of inequalities 40 and 43 we obtain the expression 34. This concludes the proof of the lemma.

Corollary I   *Let n be a fixed positive integer and let $h > 2$. The functions of R_0 satisfying condition 33 for some nonnegative integer*

$j < 2^n$ can be enclosed in a finite set of quasi-intervals of the collection $Q(\nu; m_1, \cdots, m_{2^\nu})$, the sum of whose probabilities is less than

$$\frac{1}{h2^n} + 2^{n+1} \exp\left[-\frac{h}{2} 2^{n(\frac{1}{2} - \lambda)} \right]. \tag{44}$$

The functions of R_0 satisfying expression 33 for at least one positive integer n and at least one nonnegative integer $j < 2^n$ can be enclosed in a denumerable number of quasi-intervals of the collection $Q(\nu; m_1, \cdots, m_{2^\nu})$ the sum of whose probabilities is less than $p(h)$, where

$$\lim_{h \to \infty} p(h) = 0. \tag{45}$$

Proof. The first part of the corollary follows at once from Lemma 1 by multiplying the bound 34 by 2^n.

For the second part we first note that if q is any positive integer, then there is a positive number x_q such that $x^q < e^x$ for $x > x_q$. If we take q a positive integer satisfying the inequality $q(\frac{1}{2} - \lambda) \geq 2$, then for N_q sufficiently large we have

$$2^{n+1} \exp\left[-\frac{h}{2} 2^{n(\frac{1}{2} - \lambda)} \right] < 2^{n+1} \cdot \frac{2^q}{h^q 2^{n(\frac{1}{2} - \lambda)q}} < 2^{n+1} \frac{2^q}{h^q} \cdot \frac{1}{2^{2n}} = \frac{2^{q+1}}{h^q} \cdot \frac{1}{2^n} \tag{46}$$

for $n > N_q$. Thus for $n > N_q$ the expression 44 is less than

$$\frac{1}{h} \frac{1}{2^n} \left(1 + \frac{2^{q+1}}{h^q} \right); \tag{47}$$

hence the infinite series obtained by summing the expression 44 from one to infinity converges, and if we denote its sum by $p(h)$, we have the statement 45.

We close this section by proving the following lemma:

Lemma 2   Let $0 < \lambda < \frac{1}{2}$, and let $h > 0$. If a function of R_0 satisfies the inequality

$$|x(t_2) - x(t_1)| \leq h|t_2 - t_1|^\lambda \tag{48}$$

for all "adjacent" binary points t_1, t_2 in $0 \le t \le 1$, then it satisfies the inequality

$$|x(t_2) - x(t_1)| \le \frac{2h}{1 - 2^{-\lambda}} |t_2 - t_1|^\lambda \tag{49}$$

for all binary points t_1, t_2 in $0 \le t \le 1$.

(By two "adjacent" binary points we mean two whose difference is 2^{-n} for some positive integer n.)

Before giving the proof, we consider an example that will serve to illustrate the proof. For the example we assume that inequality 48 holds for all adjacent binaries, and we seek to prove inequality 49 for the particular values $t_1 = \frac{1}{16}$, $t_2 = \frac{61}{64}$. If we write $\frac{1}{16} = \frac{4}{64}$, then an obvious way to approach the problem is to use inequality 48 repeatedly in steps of $1/64$, namely

$$|x(\tfrac{61}{64}) - x(\tfrac{4}{64})| \le |x(\tfrac{61}{64}) - x(\tfrac{60}{64})|$$
$$+ |x(\tfrac{60}{64}) - x(\tfrac{59}{64})| + \cdots + |x(\tfrac{5}{64}) - x(\tfrac{4}{64})|$$
$$\le 57h(\tfrac{1}{64})^\lambda = (57)^{1-\lambda}h(\tfrac{57}{64})^\lambda.$$

For values of λ sufficiently near $\frac{1}{2}$ but $<\frac{1}{2}$, the factor $(57)^{1-\lambda}$ is greater than the factor $2/(1 - 2^{-\lambda})$ which occurs in inequality 49, and hence this estimate is too crude. Accordingly we try a slight modification of this method. We introduce a properly chosen intermediate point t_3 between t_1 and t_2, and obtain estimates on $|x(t_3) - x(t_1)|$ and on $|x(t_2) - x(t_3)|$. It turns out that a helpful choice of t_3 is the value $t_3 = \frac{1}{2}$. (We shall later discuss the basis of this choice.) We now write

$$t_3 - t_1 = \tfrac{1}{2} - \tfrac{1}{16} = \tfrac{7}{16} = \tfrac{1}{4} + \tfrac{1}{8} + \tfrac{1}{16}$$
$$t_2 - t_3 = \tfrac{61}{64} - \tfrac{1}{2} = \tfrac{29}{64} = \tfrac{1}{4} + \tfrac{1}{8} + \tfrac{1}{16} + \tfrac{1}{64}$$

and again use inequality 48, but this time as follows:

$$|x(t_3) - x(t_1)| \le |x(\tfrac{1}{2}) - x(\tfrac{1}{4})| + |x(\tfrac{1}{4}) - x(\tfrac{1}{8})| + |x(\tfrac{1}{8}) - x(\tfrac{1}{16})|$$
$$\le h\left[\frac{1}{4^\lambda} + \frac{1}{8^\lambda} + \frac{1}{16^\lambda}\right]$$
$$= h \cdot \frac{1}{4^\lambda}\left[1 + \frac{1}{2^\lambda} + \frac{1}{2^{2\lambda}}\right] < \frac{h}{1 - 2^{-\lambda}} \cdot \frac{1}{4^\lambda}$$

and

$$|x(t_2) - x(t_3)| \le |x(\tfrac{1}{2}) - x(\tfrac{1}{2} + \tfrac{1}{4})| + |x(\tfrac{1}{2} + \tfrac{1}{4})$$
$$- x(\tfrac{1}{2} + \tfrac{1}{4} + \tfrac{1}{8})| + \cdots + |x(\tfrac{1}{2} + \tfrac{1}{4} + \tfrac{1}{8} + \tfrac{1}{16})$$
$$- x(\tfrac{1}{2} + \tfrac{1}{4} + \tfrac{1}{8} + \tfrac{1}{16} + \tfrac{1}{64})|$$
$$\le h\left[\frac{1}{4^\lambda} + \frac{1}{8^\lambda} + \frac{1}{16^\lambda} + \frac{1}{64^\lambda}\right]$$
$$\le \frac{h}{1 - 2^{-\lambda}} \cdot \frac{1}{4^\lambda}.$$

On combining these two inequalities and noting that $t_2 - t_1 \ge \tfrac{1}{4}$, we obtain inequality 49 for $t_1 = \tfrac{1}{16}$ and $t_2 = \tfrac{61}{64}$.

We now turn to the proof of Lemma 2 in general. Let t_1 and $t_2 > t_1$ be two binary numbers in $(0, 1)$ and let us represent them as binary fractions

$$t_1 = 0.a_1 a_2 \cdots a_n \cdots, \quad t_2 = 0.b_1 b_2 \cdots b_n \cdots.$$

Let t_3 be a number whose binary expansion may be made to agree with t_1 up to and including a_j, and with that of t_2 up to and including b_j. We choose t_3 so that j is as large as possible, even though this may necessitate the use of an expression for t_3 ending in $111 \cdots$ to agree with t_1 and of a terminating expression for t_3 to agree with t_2. (In the example

$$t_1 = \tfrac{1}{16} = 0.00010000 \cdots, \quad t_2 = \tfrac{61}{64} = 0.111101000 \cdots.$$

On writing

$$t_3 = 0.01111111 \cdots = 0.1000 \cdots (= \tfrac{1}{2})$$

we see that $j = 1$; i.e., the first representation agrees with that of t_1 for $j = 1$, and the second with that of t_2 for $j = 1$.)

With j so defined, we have

$$t_2 - t_1 \ge \frac{1}{2^{j+1}} \tag{50}$$

and we also see that the differences $t_3 - t_1$ and $t_2 - t_1$ are expressible in the forms

$$t_3 - t_1 = 0.0 \cdots 0c_{j+1}c_{j+2}\cdots, \quad t_2 - t_3 = 0.0 \cdots 0d_{j+1}d_{j+2}\cdots \tag{51}$$

where there are j consecutive zeros after the point in each, and every c and d is zero or one. We write $\alpha_0 = 0$ and

$$\alpha_s = 0.0\cdots0c_{j+1}c_{j+2}\cdots c_{j+s} \qquad (52)$$

for $s = 1, 2, \cdots$. We can then go from t_1 to t_3 by successive steps consisting of going from one binary point to an "adjacent" one, as follows:

$$t_1 - t_3 = \sum_{s=0}^{\infty} [(t_1 + \alpha_s) - (t_1 + \alpha_{s+1})], \qquad (53)$$

and hence by our hypothesis, inequality 48, we have

$$|x(t_1) - x(t_3)| \le h \sum_{s=0}^{\infty} \frac{c_{j+s+1}}{2^{(j+s+1)\lambda}} \le \frac{h}{2^{(j+1)\lambda}} \sum_{s=0}^{\infty} \frac{1}{2^{s\lambda}}$$
$$= \frac{h}{1 - 2^{-\lambda}} \cdot \frac{1}{2^{(j+1)\lambda}}. \qquad (54)$$

Similarly,

$$|x(t_3) - x(t_2)| \le \frac{h}{1 - 2^{-\lambda}} \cdot \frac{1}{2^{(j+1)\lambda}}. \qquad (55)$$

Using inequality 50, we see that inequalities 54 and 55 yield the lemma.

Using Corollary 1 and Lemma 2, we have Theorem 1.

4. The nonbounded variation property

In this section we shall show that almost no function of the Brownian motion process is of bounded variation, a result first proved by Wiener [1923]. Following Lévy [1940], we shall prove a little more than this.

Let m be a positive integer, and let

$$0 \le t_0 < t_1 < \cdots < t_m \le 1. \qquad (56)$$

Denote the increments and time differences by

$$\Delta X_j = X(t_j, \alpha) - X(t_{j-1}, \alpha); \Delta t_j = t_j - t_{j-1}; \qquad j = 1, \cdots, m \ (57)$$

and the corresponding sums of squares by

$$B_m = \sum_{j=1}^{m} (\Delta X_j)^2, \qquad b_m = \sum_{j=1}^{m} (\Delta t_j)^2. \qquad (58)$$

Recalling that ΔX_j is a Gaussian variable with mean zero and variance Δt_j, and that the increments ΔX_j and ΔX_p are independent for $j \neq p$, we see that

$$
\begin{aligned}
E\{[(\Delta X_j)^2 - \Delta t_j]^2\} &= E\{(\Delta X_j)^4\} - 2E\{(\Delta X_j)^2\}\Delta t_j + (\Delta t_j)^2 \\
&= 3(\Delta t_j)^2 - 2(\Delta t_j)^2 + (\Delta t_j)^2 \\
&= 2(\Delta t_j)^2,
\end{aligned}
\tag{59}
$$

and

$$
E\{[(\Delta X_j)^2 - \Delta t_j][(\Delta X_p)^2 - \Delta t_p]\} = 0; \qquad j \neq p, \tag{60}
$$

where in Equation 59 we have used the relation

$$
E\{(\Delta X_j)^4\} = \frac{(\Delta t_j)^2}{\sqrt{2\pi}} \int_{-\infty}^{\infty} u^4 e^{-u^2/2} \, du = 3(\Delta t_j)^2. \tag{61}
$$

On using Equations 59 and 60, we see that

$$
\begin{aligned}
E[B_m - (t_m - t_0)]^2 &= E\left\{\left[\sum_{j=1}^{m} (\Delta X_j)^2 - \Delta t_j\right]^2\right\} \\
&= \sum_{j=1}^{m} E\{[(\Delta X_j)^2 - \Delta t_j]^2\} \\
&= 2 \sum_{j=1}^{m} (\Delta t_j)^2 \leq 2(t_m - t_0) \cdot \max_{j=1,\cdots,m} \Delta t_j.
\end{aligned}
\tag{62}
$$

On choosing m to be of the form 2^n, and $t_j = j/2^n$, we shall conclude the following lemma from Equation 62.

Lemma 3   *Denote by $\sigma_n(\alpha)$ the sum*

$$
\sigma_n(\alpha) = \sum_{j=1}^{2^n} \left[X\left(\frac{j}{2^n}, \alpha\right) - X\left(\frac{j-1}{2^n}, \alpha\right) \right]^2. \tag{63}
$$

Then $\sigma_n(\alpha)$ converges to one in the mean,

$$
\lim_{n \to \infty} E\{[\sigma_n(\alpha) - 1]^2\} = 0, \tag{64}
$$

and for almost all α the limit of $\sigma_n(\alpha)$ exists and is one,

$$
\lim_{n \to \infty} \sigma_n(\alpha) = 1. \tag{65}
$$

The first relation, 64, follows at once from Equation 62. To prove the second relation, 65, for almost all α, we continue to follow the proof given by Lévy [1940]. For m and t_j so defined, the inequality 62 yields

$$E\{[\sigma_n(\alpha) - 1]^2\} \leq 2^{-n+1}; \qquad n = 1, 2, \cdots, \tag{66}$$

and this implies that

$$m\{\alpha: |\sigma_n(\alpha) - 1| \geq n2^{-n/2}\} \leq \frac{2}{n^2}; \qquad n = 1, 2, \cdots. \tag{67}$$

Hence for each positive integer n

$$|\sigma_j(\alpha) - 1| < j2^{-j/2}; \qquad j = n, n + 1, \cdots, \tag{68}$$

except for at most an α-set of measure

$$2\left(\frac{1}{n^2} + \frac{1}{(n + 1)^2} + \cdots\right). \tag{69}$$

Since $j2^{-j/2} \to 0$ as $j \to \infty$, and since the series $\sum 1/n^2$ converges, this yields statement 65 almost everywhere. This concludes the proof of Lemma 3.

Lévy [1940] has proved a more general result of the same nature, and Doob [1953] has given an alternative proof of this more general result. We shall merely state Doob's version of the result as a lemma, but shall not give the proof, and we do not need to make use of the result.

Lemma 4   *Let $t_0, t_1, \cdots$ be everywhere dense in the interval $(0, 1)$ and let $t_0^{(n)}, \cdots, t_n^{(n)}$ be the numbers $t_0, \cdots, t_n$ arranged in numerical order, $t_0^{(n)} < \cdots < t_n^{(n)}$. Then*

$$\lim_{n \to \infty} \sum_{j=1}^{n} [X(t_j^{(n)}, \alpha) - X(t_{j-1}^{(n)}, \alpha)]^2 = 1 \tag{70}$$

with probability one, and the limit relation is also true in the sense of convergence in the mean.

The conclusion about convergence in the mean follows, of course, directly from Equation 62.   The proof of the limit relation 70 would require further considerations.

Having Lemma 3, we shall prove the following result.

Theorem 2   *Almost no sample function $X(t, \alpha)$ of the Brownian motion process is of bounded variation.*

Proof. Let $f(t)$ be any fixed continuous function on $(0, 1)$. It follows that

$$\sum_{j=1}^{2^n} \left[f\left(\frac{j}{2^n}\right) - f\left(\frac{j-1}{2^n}\right) \right]^2$$

$$\leq \max_{1 \leq j \leq 2^n} \left| f\left(\frac{j}{2^n}\right) - f\left(\frac{j-1}{2^n}\right) \right| \cdot \sum_{j=1}^{2^n} \left| f\left(\frac{j}{2^n}\right) - f\left(\frac{j-1}{2^n}\right) \right|. \quad (71)$$

If the sum on the right were bounded independently of n (which it would be if f were of bounded variation), then the right-hand side would go to zero as n approaches infinity. Thus Lemma 3 implies Theorem 2.

In Section 3 we showed that for each $\lambda < \frac{1}{2}$ almost all sample functions $X(t, \alpha)$ of the Brownian motion process satisfy an inequality of the form

$$|X(t_2, \alpha) - X(t_1, \alpha)| < h|t_2 - t_1|^\lambda \quad (72)$$

for some h and all values of t_1, t_2 in $(0, 1)$. This same property cannot hold for any $\lambda > \frac{1}{2}$. In fact, we can say a little more, namely:

Corollary 2   *Let $\lambda > \frac{1}{2}$ be a fixed number. Then the set S of sample functions $X(t, \alpha)$ of the Brownian motion process for which an inequality of the form given in 72 holds for some h and for all binary points t_1, t_2 in $(0, 1)$ has measure zero.*

Proof. If $X(t, \alpha) \in S$, then by hypothesis there is a positive number h (depending in general upon α) such that

$$\sum_{j=1}^{2^n} \left[X\left(\frac{j}{2^n}, \alpha\right) - X\left(\frac{j-1}{2^n}, \alpha\right) \right]^2 \leq 2^n \cdot h \cdot 2^{-2n} = h 2^{n(1-2\lambda)}. \quad (73)$$

Since the right-hand side approaches zero as n approaches infinity, the relation 65, which holds for almost all α, implies that $m(S) = 0$.

If a function is of bounded variation, or if it satisfies a uniform Lipschitz condition of order $\lambda > \frac{1}{2}$ on $0 \leq t \leq 1$, then, as we have seen, the value of the limit in statement 65 is zero. Loud [1954], pointing out that for most "ordinary" functions the value of this limit is zero, constructed an example of a function $x_0(t)$, continuous on $(0, 1)$ and vanishing at $t = 0$, for which the limit is actually one. His example is the function

$$x_0(t) = \sqrt{2} \sum_{1}^{\infty} 2^{-m/2} g(t, 2^{-m}),$$

where $g(t, \rho)$ is any function which is periodic of period 2ρ, equal to zero for even multiples of ρ, equal to one for odd multiples of ρ, and linear between.

Lévy [1948, pp. 247–248] has proved that for each fixed t_0,

$$\limsup_{t \to t_0} \frac{X(t, \alpha) - X(t_0, \alpha)}{\sqrt{2|t - t_0| \log(1/|t - t_0|)}} = 1 \tag{74}$$

holds with probability 1, and this implies in particular that

$$\limsup_{t \to t_0} \frac{X(t, \alpha) - X(t_0, \alpha)}{t - t_0} = \infty \tag{75}$$

with probability 1. As Doob [1953, p. 394] points out, this means that almost all sample functions have the property that the upper derivates are $+\infty$ for all values of t, except for a set of values of t of Lebesgue measure 0, the exceptional set varying with the sample function.

Physically it is reasonable to require that the sample functions of the process be continuous and this gives strength to the adoption of the Brownian motion process as the model of the physical Brownian motion. There is, however, the serious difficulty that these sample functions are not of bounded variation so that the paths of the particles (in the model) have infinite length, and the velocities are infinite. Doob [1953, p. 398; 1942], and others have discussed these disadvantages, and, as mentioned in footnote 1, Uhlenbeck and Ornstein [1930] have defined a different stochastic process describing the Brownian motion. For the present we shall give the following quotation from Doob [1953, p. 398]:

These properties of the [Brownian motion] process should not be taken too seriously from a practical point of view since they are *in the small* properties of the sample functions, involving increments $x(t + s) - x(t)$ with s small, and we have already remarked that the fit of theory to practice cannot be expected to be good for properties involving small increments.

The Brownian motion process occurs in many places in the theory of stochastic processes, and it has various physical applications. We shall briefly discuss several of these aspects of the process later in this chapter.

5. Generality of the Brownian process from the mathematical point of view

Let $X_1, X_2, \cdots$, be independent random variables each with mean 0 and variance 1 so that the central limit theorem is applicable. Denote by S_n the nth cumulative sum

$$S_n = X_1 + \cdots + X_n. \tag{76}$$

Let $V(x)$ be a nonnegative function of the real variable x in $(-\infty, \infty)$. Kac [1949] has shown that under very general conditions on V, the limiting distributions of the sums

$$\frac{1}{n} \sum_{j \leq nt} V\left(\frac{S_j}{\sqrt{n}}\right) \tag{77}$$

can be obtained by calculating the probability of a suitable functional of the Brownian motion process, namely

$$\lim_{n \to \infty} P\left\{\frac{1}{n} \sum_{j \leq nt} V\left(\frac{S_j}{\sqrt{n}}\right) < \beta\right\} = P\left\{\int_0^t V[X(\tau, \alpha)] \, d\tau < \beta\right\} \tag{78}$$

for $t > 0$, where $\{X(t, \alpha), 0 \leq t < \infty\}$ is a Brownian motion process. This result, as well as various results leading up to it, and a more general result of the same nature due to Donsker [1951], will be described in considerable detail in the final sections of Chapter 3.

Another basic mathematical property of the Brownian motion process is related to the fact that the sample functions of the process are almost all continuous functions: Any process with independent increments whose sample functions are almost all continuous is essentially the Brownian motion process. A full discussion of this fact can be found in Doob [1953, Section 7, Chapter 8] and Lévy [1948].

6. Relation to noise and information theory

The foundation of the theory of noise in electrical circuits is Nyquist's theorem, according to which the power spectrum (power per unit frequency range as a function of frequency) of the noise voltage in a resistor has the constant value RT/N. The shot effect due to spontaneous fluctuations in the intensity of the stream of electrons flowing from the cathode to the anode in a vacuum tube is a typical example of noise. When it is further assumed that the noise voltage is a random function negligibly correlated with itself between any two time instants, the result is a function that may be treated in formal manipulations as if it were the "derivative" of a sample function of the Brownian motion process.

(Laning and Battin [1956] have described this relation as follows. Denoting a random process $\{x_t\}$ possessing a constant power spectral density as a *white noise*, they visualize the physical model of a white noise as a collection of small random impulses, occurring at a high rate of frequency, and they show that this process and the Brownian motion process are related formally by the equation $X(t) = \int_0^t x_\tau \, d\tau$.)

(Fortet [1958, p. 198] discusses this point as follows: "As is well known, we cannot ascribe a derivative in one sense or another to $X(t, \alpha)$ considered as a random *function*. Nevertheless, technicians often employ a random function $Y(t)$, called 'white noise,' as the derivative of $X(t, \alpha)$, and their computations are valid in practice although they are not based on a correct definition of the differentiability of $X(t, \alpha)$." Fortet then goes on to discuss the use of *Schwartz distributions*.)

Further developments and refinements in the theory of noise invariably use this random function as an implicit base. The applications of noise theory involve mainly the question of extracting the optimum amount of information from an electrical signal contaminated by noise introduced in the atmosphere, in the circuits themselves, or from other random sources.

Discussion of this theory may be found in the papers by Rice [1944, 1945], the books by Uhlenbeck and Lawson [1950] and Laning and Battin [1956], and in the references cited in these places. As various authors have pointed out, functionals of the type

$$\int_0^t V[X(\tau, \alpha)] \, d\tau \qquad\qquad (79)$$

which occur on the right-hand side of Equation 78 are not only of great interest for a theoretical study of $X(t, \alpha)$, but they are also of interest for practical reasons. For example, if

$$V(x) = \begin{cases} 0, & \text{for } x \le a, \\ 1, & \text{for } a < x, \end{cases}$$

then as Fortet [1958, p. 204] points out, in a communication problem in which $X(t, \alpha)$ represents noise, the functional 79 gives the relative time that noise does not exceed a certain level. Continuing this discussion, Fortet has written:

In communication theory, it often happens that the noise $X(t)$ goes through a nonlinear device (e.g., a detector or a rectifier) which transforms $X(t, \alpha)$ into $V[X(t, \alpha)]$. For example, $V(x) = |x|$ or $V(x) = x^2$. Following this, there may be an integrating device, resulting in

$$L(t, \alpha) = \int_{t-\tau}^t V[X(u, \alpha)] \, du.$$

Then we have the problem of determining the statistical properties of $L(t, \alpha)$ from those of $X(t, \alpha)$.

In Sections 4 and 5 of Chapter 3 we discuss the distribution of the functional 79 for various special cases of $V(x)$, in particular for $V(x) = |x|$ and $V(x) = x^2$, and in Section 7 of Chapter 3 we consider the work of Kac [1949] on the distribution of the functional in the general case.

7. Relation to heat conduction

The molecular mechanism of heat conduction may be reasonably assumed to have a strong similarity to diffusion. Since there is not an actual migrating particle, however, it is most convenient to derive the relevant partial differential equation from phenomenological considerations. The basic physical assumptions are that the rate of heat flow at any point of a medium is proportional to the temperature gradient, and that the temperature increment of any small thermally homogeneous region in a substance is proportional to the net flux of heat into it (barring other kinds of energy transfer). With the use of Gauss's theorem concerning vector fields, it then follows that the temperature in a medium as a function of position vector $\vec{r}$ and time t satisfies the equation

$$\frac{\partial T}{\partial t} = \mu \nabla^2 T. \tag{80}$$

We might have derived the diffusion equation 3 in an exactly similar fashion from the assumption that the net flow of solute particles at any point of a medium is proportional to the concentration gradient. This assumption may be regarded as a phenomenological consequence of the random processes that are described in more detailed fashion in considerations of particle motion as a random process.

8. Other connections

In his paper on "Stochastic Processes in Physics and Astronomy," Chandrasekhar [1943] discusses among other matters several physical phenomena in which the theory of Brownian motion enters. Among these are the problem of the random walk, the Langevin Equation, and the Fokker–Planck Equation, the equation of diffusion (or of heat conduction), the transition between the macroscopically irreversible nature of diffusion and the microscopically reversible nature of molecular fluctuations, the phenomenon of coagulation exhibited by colloidal particles when an electrolyte is added to the solution, and problems on stellar dynamics. In this paper Chandrasekhar also develops in considerable detail the mathematical theory of the Brownian motion (in three dimensions), and he gives many references to the work of others. Other references which we would like to mention explicitly are the papers by Uhlenbeck and Ornstein [1930],

Wang and Uhlenbeck [1945], and Kac [1947]. As we mentioned in Section 4, more exact physical treatments of the Brownian motion lead to a different mathematical model, yielding a stochastic process whose sample functions are not only continuous but also have continuous first derivatives. This theory was developed by Uhlenbeck and Ornstein [1930] and investigated further by Chandrasekhar [1943], Doob [1942], and Wang and Uhlenbeck [1945].

3

Integration in Differential Space

In the first section of this chapter we define integrals of the form $\int_T \phi(t) \, dX(t, \alpha)$, where $\phi \in L_2$ over T, and $X(t, \alpha)$ is the sample function of a (separable) Brownian motion process defined for $t \in T$, where T is a fixed one of the three intervals $[0, 1]$, $[0, \infty)$, or $(-\infty, \infty)$. This will be followed in Sections 2 and 3 with expansions of functions $F(\alpha)$ of L_2 over T in terms of functions involving the random variables $\int_T \phi_\mu(t) \, dX(t, \alpha)$, where $\{\phi_\mu(t)\}$ is a complete orthonormal set on $L_2(T)$.

In Sections 2 and 3 the interval T can be any one of the three intervals just listed, while in Sections 4 through 7 it will ordinarily be the interval $[0, \infty)$. The work in Sections 2 and 3 depends on the work of Section 1, but Sections 4 through 7 are independent of the earlier parts of this chapter.

In Section 4 we list the distribution functions for a set of selected functionals, and we briefly discuss interpretations of these. In the next section we describe an invariance principle, or limiting distribution, related to the formulas of Section 4. In Section 6 we continue the discussion of Sections 4 and 5, and give the distribution functions of certain other functionals, as well as some additional instances of the invariance principle. We close Section 6 with a statement of three asymptotic theorems of a different type, two due to Kallianpur and Robbins [1953], and the third a generalization of one of these, due to Derman [1954]. We close the chapter with a description of two general formulations of the invariance principle, due to Kac [1949] and to Donsker [1951]. In the last four sections, the results are merely described without proofs.

1. Integration with respect to $X(t, \alpha)$

In this section, $X(t, \alpha)$ will be the sample function of the Brownian motion process defined for $t \in T$, where T is a fixed one of the three intervals $[0, 1]$, $[0, \infty)$, or $(-\infty, \infty)$.

There is a natural relation between the Brownian motion process and Hilbert space. A hint of the existence of such a relation lies in the theorem of Rademacher [1922] which asserts that if real numbers c_n are such that

$$\sum_{-\infty}^{\infty} c_n^2 < \infty,$$

then, for almost all choices of $+$ or $-$, $\sum_{-\infty}^{\infty} \pm c_n$ is convergent. There is a continuous version of this phenomenon in which the summation is replaced by integration, and $\pm$ by the "differential" of the Brownian motion.

We shall define integrals of the form

$$\Phi(\alpha) = \int_T \phi(t) \, dX(t, \alpha), \qquad (1)$$

where ϕ belongs to L_2 over T. Since $X(t, \alpha)$ is almost never of bounded variation, this integral cannot in general be defined as an ordinary Stieltjes integral. Wiener [1923] has introduced integrals of this form, and has given the integral 1 an interpretation as a generalized Stieltjes integral. His work has been extended to include integrals with respect to any stochastic process $\{Y_t, t \in T\}$ with orthogonal increments (see Doob [1953]). In this section we shall define the integral 1 for the Brownian motion process. Our presentations will follow that given by Doob, but with the stochastic process $\{Y_t, t \in T\}$ specialized to be the Brownian motion process.

We may say that a step function $s(t)$ defined on T is of type (f) if there are a positive integer n and $2n + 1$ numbers $t_1, \cdots, t_{n+1}$, $c_1, \cdots, c_n$ with

$$t_1 < t_2 < \cdots < t_{n+1}, t_j \in T; \qquad j = 1, \cdots, n + 1 \qquad (2)$$

such that

$$s(t) = \begin{cases} 0, & \text{for} & t < t_1, \\ c_j, & \text{for} & t_j \leq t < t_{j+1}, & j = 1, \cdots, n, \quad (3) \\ 0, & \text{for} & t_{n+1} < t. \end{cases}$$

We first define the integral 1 when ϕ is a step function of type (f). If $\phi(t)$ is the step function 3, then we define the integral 1 as

$$\sum_{j=1}^{n} c_j[X(t_{j+1}, \alpha) - X(t_j, \alpha)]. \quad (4)$$

For ϕ thus specialized we see that

$$E\{\Phi(\alpha)\} = 0 \quad (5)$$

and

$$E\{|\Phi(\alpha)|^2\} = E\{|\sum c_j[X(t_{j+1}, \alpha) - X(t_j, \alpha)]|^2\}$$
$$= \sum_{j, k=1}^{n} c_j \bar{c}_k E\{[X(t_{j+1}, \alpha) - X(t_j, \alpha)][X(t_{k+1}, \alpha)$$
$$- X(t_k, \alpha)]\}$$
$$= \sum_{j=1}^{n} |c_j|^2(t_{j+1} - t_j) = \int_T |\phi(t)|^2 \, dt. \quad (6)$$

Similarly, if ψ is a step function of type (f), and if $\Psi(\alpha)$ is the integral

$$\Psi(\alpha) = \int_{-\infty}^{\infty} \psi(t) \, dX(t, \alpha), \quad (7)$$

then

$$E\{\Phi(\alpha)\overline{\Psi(\alpha)}\} = \int_T \phi(t)\overline{\psi(t)} \, dt. \quad (8)$$

Since the step functions of type (f) are everywhere dense in $L_2(T)$, we can define the integral 1 for every function ϕ belonging to L_2 over T in the following manner. Let $\phi \in L_2$ and let $\{\phi_\mu(t)\}$ be a sequence of step functions, each of type (f), and such that

$$\underset{\mu \to \infty}{\text{l.i.m.}} \, \phi_\mu(t) = \phi(t). \quad (9)$$

Define the integral $\Phi_\mu(\alpha)$ for each function $\phi_\mu(t)$, $\mu = 1, 2, \cdots$. By Equation 6

$$E\{|\Phi_\mu(\alpha) - \Phi_\nu(\alpha)|^2\} = \int_T |\phi_\mu(t) - \phi_\nu(t)|^2 \, dt, \quad (10)$$

and thus the sequence $\{\Phi_\mu(\alpha)\}$ is a convergent sequence in the mean-square sense. The Riesz-Fischer theorem tells us that there exists a function $\Phi(\alpha)$ of L_2 such that

$$\lim_{\mu \to \infty} E\{|\Phi(\alpha) - \Phi_\mu(\alpha)|^2\} = 0. \tag{11}$$

We also have

$$E\{|\Phi(\alpha)|^2\} = \int_T |\phi(t)|^2 \, dt. \tag{12}$$

Moreover the limit in the mean $\Phi(\alpha)$ does not depend upon the sequence $\{\phi_\mu(t)\}$ that approximates $\phi(t)$, but will be the same almost everywhere for any sequence. That is, if l.i.m. $\psi_\mu(t) = \phi(t)$, and if each $\psi_\mu(t)$ is a step function of type (f), then the integrals $\Psi_\mu(\alpha)$ defined in terms of $\psi_\mu(t)$ converge in the mean to $\Phi(\alpha)$. One sees this, for example, by defining a new sequence $\{\eta_\mu(t)\}$ with $\eta_{2\mu}(t) = \phi_\mu(t)$, $\eta_{2\mu+1}(t) = \psi_\mu(t)$. We have l.i.m. $\eta_\mu(t) = \phi(t)$, and hence the random variable $\int_T \eta_\mu(t) \, dX(t, \alpha)$ converges in the mean to some function. But this function must be $\Phi(\alpha)$ almost everywhere.

This gives the means of defining the integral 1 for $\phi(t)$ any function belonging to $L_2(T)$. Thus defined, it is a unique random variable; it exists for almost all values of α, and the relations 5, 8, and 12 hold. It also follows that the Riemann-Stieltjes sum for the integral 1 converges in the mean whenever $\phi(t)$ is Riemann integrable, and that the usual rules for manipulating integrals are valid. In particular, integration by parts is valid if ϕ is of bounded variation.

We shall prove the following theorem due to Paley, Wiener, and Zygmund [1933]:

Theorem 1   *If ϕ is a real-valued function belonging to $L_2(T)$, then the integral 1 is a Gaussian variable with mean zero and variance $\int_T \phi^2(t) \, dt$. If $\{\phi_\mu(t)\}$ is an orthogonal set of real functions, each belonging to $L_2(T)$, then the random variables*

$$\Phi_\mu(\alpha) = \int_T \phi_\mu(t) \, dX(t, \alpha); \qquad \mu = 1, 2, \cdots, \tag{13}$$

are (statistically) independent.

Proof. We have already seen that the integral 1 considered as a random variable has mean zero and variance $\int_T \phi^2(t)\, dt$. For the first part of the theorem it remains only to show that the integral 1 is Gaussian. For this purpose let us first note that if ϕ is a step function of type (f), then the linear combination given in 4 which defines the integral is Gaussian. In the general case if $\{\phi_\mu(t)\}$ is a sequence of step functions of type (f) which approximate ϕ in the L_2-sense, then each of the integrals $\Phi_\mu(\alpha)$ is Gaussian, and as we have seen in statement 11, $\Phi_\mu(\alpha)$ approaches $\Phi(\alpha)$ in the L_2-sense, and the variance $\int_T [\phi_\mu(t)]^2\, dt$ of Φ_μ approaches the variance of Φ. Thus Φ itself is Gaussian. This yields the first part of the theorem.

Before proving the second part of the theorem, the independence of the Φ_μ, we outline the essential ideas. First we shall show that for any positive integer m, the m random variables

$$\Phi_1, \cdots, \Phi_m \tag{14}$$

have a multivariate Gaussian distribution. Since they are known to be uncorrelated,

$$E\{\Phi_\mu \Phi_\nu\} = 0; \qquad \text{for } \mu \neq \nu, \tag{15}$$

this implies their independence.

To show that the random variables in the set 14 have a multivariate Gaussian distribution, let

$$\{\phi_{\mu,p}; \qquad p = 1, 2, \cdots\} \tag{16}$$

be a sequence of step functions of type (f) which converge in the mean to ϕ_μ,

$$\underset{p \to \infty}{\text{l.i.m.}}\ \phi_{\mu,p}(t) = \phi_\mu(t) \tag{17}$$

and let $\Phi_{\mu,p}(\alpha)$ be the associated random variable defined as was the integral 1. Then for each index $p = 1, 2, \cdots$ the m functions

$$\Phi_{1,p}, \cdots, \Phi_{m,p} \tag{18}$$

have a multivariate Gaussian distribution since each is a (finite) linear combination of increments of $X(t, \alpha)$ as in expression 4.

Hence the m functions 14 have a multivariate Gaussian distribution. (The characteristic function for the set of functions is given by

$$f_p(v_1, \cdots, v_m) = \exp\left\{ -\tfrac{1}{2} \sum_{\mu=1}^{m} \sum_{\nu-1}^{m} \rho_{\mu,\nu}^{(p)} v_\mu v_\nu \right\}, \tag{19}$$

where $\rho_{\mu,\nu}^{(p)}$ is the correlation of $\Phi_\mu^{(p)}$, $\Phi_\nu^{(p)}$, namely

$$\rho_{\mu,\nu}^{(p)} = E\{\Phi_\mu^{(p)} \Phi_\nu^{(p)}\}. \tag{20}$$

Clearly

$$\lim_{p \to \infty} \rho_{\mu,\nu}^{(p)} = E\{\Phi_\mu(\alpha)\Phi_\nu(\alpha)\} = \begin{cases} 0, & \text{if} \quad \mu \neq \nu, \\ \int_T \phi_\mu^2(t)\, dt, & \text{if} \quad \mu = \nu, \end{cases} \tag{21}$$

and hence the characteristic function $f_p(v_1, \cdots, v_m)$ converges to the characteristic function

$$f(v_1, \cdots, v_m) = \exp\left\{ -\tfrac{1}{2}\left[v_1^2 \int_T \phi_1^2(t)\, dt + \cdots + v_m^2 \int_T \phi_m^2(t)\, dt \right] \right\}, \tag{22}$$

which is the characteristic function of m independent Gaussian variables with mean zero and variances $\int_T \phi_1^2(t)\, dt, \cdots, \int_T \phi_m^2(t)\, dt$.)

As a corollary to Theorem 1 we have the following useful formula:

Corollary I   *Let $\{\phi_\mu(t)\}$ be an orthonormal set of real functions each belonging to $L_2(T)$, and let $G(u_1, \cdots, u_p)$ be any function for which $G(u_1, \cdots, u_p) \exp\left[-(u_1^2 + \cdots + u_p^2)/2\right]$ is summable in $-\infty < u_k < \infty$; $k = 1, 2, \cdots, p$. Then*

$$E\left\{ G\left[\int_T \phi_1(t)\, dX(t, \alpha), \cdots, \int_T \phi_p(t)\, dX(t, \alpha) \right] \right\}$$

$$= (2\pi)^{-n/2} \int_{-\infty}^{\infty} \cdots \int_{-\infty}^{\infty} G(u_1, \cdots, u_p) \exp\left[-\frac{(u_1^2 + \cdots + u_p^2)}{2} \right]$$

$$\times\, du_1 \cdots du_p. \tag{23}$$

We close the present section with two remarks which are of general interest but which refer to ideas that we will not be using in this book.

Remark on Theorem 1. If $\{\phi_\mu(t)\}$ is a complete orthonormal set of real functions on $L_2(T)$, then the integrals

$$\Phi_\mu(\alpha) = \int_T \phi_\mu(t)\, dX(t, \alpha) \qquad (24)$$

would be the Fourier coefficients of the derivative $dX(t, \alpha)/dt$ if this derivative existed and belonged to L_2 over T. In this case, we would also have convergence of the series $\sum_1^\infty [\Phi_\mu(\alpha)]^2$, indeed to the sum $\int_T [dX(t, \alpha)/dt]^2\, dt$. But as we saw in the preceding chapter, the sample function $X(t, \alpha)$ of the Brownian motion process is almost never differentiable, and as we have just seen in Theorem 1, the integrals 24 are independent random variables, each having a Gaussian distribution with mean zero and variance one. Thus, on the average, $[\Phi_\mu(\alpha)]^2$ has the value one, and

$$E\left\{ \sum_1^m [\Phi_\mu(\alpha)]^2 \right\} = m,$$

so that, on the average, the series $\sum_1^\infty [\Phi_\mu(\alpha)]^2$ diverges (in fact, diverges with probability one).

Remark on more general stochastic integrals. Itō [1944] has defined integrals of the form

$$\int_T \phi(t, \alpha)\, dX(t, \alpha),$$

where ϕ is a suitably restricted function which depends on α as well as on t. Doob [1953] has given a still more general presentation in which the integration is with respect to sample functions of a process with orthogonal increments. As an example referring to the Brownian motion process, Doob has considered the integral

$$\int_a^b [X(t, \alpha) - X(a, \alpha)]\, dX(t, \alpha); \qquad b > a,$$

and has shown that it has the evaluation

$$\tfrac{1}{2}[X(b, \alpha) - X(a, \alpha)]^2 - \tfrac{1}{2}(b - a)$$

for almost all α. At first glance the term $\frac{1}{2}(b - a)$ is unexpected. It arises as the value of the following limit in the mean,

$$\frac{1}{2} \underset{\delta \to 0}{\text{l.i.m.}} \sum_{j=0}^{n-1} [X(t_{j+1}, \alpha) - X(t_j, \alpha)]^2,$$

where $\delta = \max_j (t_{j+1} - t_j)$. (See Section 4 of Chapter 2.) This evaluation shows for example that

$$E\left\{ \int_a^b [X(t, \alpha) - X(a, \alpha)] \, dX(t, \alpha) \right\} = 0$$

and

$$E\left\{ \left[\int_a^b (X(t, \alpha) - X(a, \alpha)) \, dX(t, \alpha) \right]^2 \right\} = \frac{1}{2}(b - a)^2.$$

2. An everywhere dense set

In this section we again consider the Brownian motion process $\{X(t, \alpha); t \in T\}$, where T is a fixed one of the three intervals $[0, 1]$, $[0, \infty)$, or $(-\infty, \infty)$. We mention explicitly that we consider the process normalized by the condition $X(0, \alpha) = 0$ for almost all α.

Let $\tau \neq 0$ be a fixed point in T. Cameron and Martin [1947] have shown that $X(\tau, \alpha)$ can be developed in a series in terms of the integrals $\int_T \phi_\mu(t) \, dX(t, \alpha)$ where $\{\phi_\mu(t)\}$ is a complete orthonormal sequence of real functions each belonging to L_2 over T. They have shown that if

$$a_\mu = E\left\{ X(\tau, \alpha) \int_T \phi_\mu(t) \, dX(t, \alpha) \right\}; \qquad \mu = 1, 2, \cdots, \quad (25)$$

then

$$\lim_{m \to \infty} E\left\{ \left[X(\tau, \alpha) - \sum_{\mu=1}^m a_\mu \int_T \phi_\mu(t) \, dX(t, \alpha) \right]^2 \right\} = 0, \quad (26)$$

and for a subsequence $m_1, m_2, \cdots$ that

$$\lim_{p \to \infty} X_{m_p}(\alpha) = X(\tau, \alpha) \quad (27)$$

for almost all α, where

$$X_m(\alpha) = \sum_{\mu=1}^m a_\mu \int_T \phi_\mu(t) \, dX(t, \alpha). \quad (28)$$

To see this, we first note that

$$X(\tau, \alpha) = \int_0^\tau dX(t, \alpha) = \int_T h(t) \, dX(t, \alpha) \tag{29}$$

for almost all α, where the function $h(t) \equiv h_\tau(t)$ is defined as

$$h(t) = \begin{cases} 1, & 0 \le t \le \tau, \\ 0, & \text{elsewhere,} \end{cases} \tag{30a}$$

if the given point τ is positive, and as

$$h(t) = \begin{cases} -1, & \tau \le t \le 0, \\ 0, & \text{elsewhere,} \end{cases} \tag{30b}$$

if τ is negative. By Equation 8 the coefficients a_μ defined in Equation 25 are also given by

$$a_\mu = \int_T h(t)\phi_\mu(t) \, dt. \tag{31}$$

Since the a_μ are the coefficients in the orthogonal development of $h(t)$, we have

$$\lim_{m \to \infty} \int_T \left[h(t) - \sum_{\mu=1}^m a_\mu \phi_\mu(t) \right]^2 dt = 0. \tag{32}$$

By Equation 12 the expected value in the limitand of Equation 26 is equal to the limitand in Equation 32 and hence Equation 26 holds. This in turn implies Equation 27.

Now this suggests that we may be able to approximate any function $F(\alpha) \in L_2(0, 1)$ by suitable combinations of the integrals $\int_T \phi_\mu(t) \, dX(t, \alpha)$. Following Cameron and Martin [1947] we shall show that this is indeed the case.

For q a positive integer, denote by S_q the class of random variables

$$f_q(\alpha) = f\left[\int_T \phi_1(t) \, dX(t, \alpha), \cdots, \int_T \phi_q(t) \, dX(t, \alpha) \right], \tag{33}$$

where $f(u_1, \cdots, u_q)$ ranges over the class of functions of q real variables such that

$$f(u_1, \cdots, u_q) \exp\left[-(u_1^2 + \cdots + u_q^2)/4\right] \in L_2(-\infty, \infty). \tag{34}$$

We denote by S the union of the S_q for $q = 1, 2, \cdots$.

We shall prove the following lemma:

Lemma 1 *The random variables of S are everywhere dense in the L_2 space over $0 \le \alpha \le 1$, that is, if $F(\alpha) \in L_2(0, 1)$, and if $\epsilon > 0$, then there exist an integer q and a function f satisfying condition 34 such that*

$$E\{|F(\alpha) - f_q(\alpha)|^2\} < \epsilon. \tag{35}$$

Proof. We shall follow the proof given by Cameron and Martin [1947]. By the usual Lebesgue argument, $F(\alpha)$ can be approximated arbitrarily closely by step functions. However, a step function is a finite linear combination of characteristic functions of α-measurable sets, and each such characteristic function can be approximated arbitrarily closely in the mean-square sense by a finite linear combination of characteristic functions of α-intervals (which are images of quasi-intervals). Thus it only remains to show that the characteristic function corresponding to a quasi-interval can be approximated arbitrarily closely by functions of class S.

For this purpose consider a quasi-interval

$$Q: \xi'_j \le X(t_j, \alpha) < \xi''_j; \qquad j = 1, \cdots, n,$$

where $t_1, \cdots, t_n$ are n fixed points in T, each different from zero. Let $\epsilon > 0$, and let $\psi_{j,\epsilon}(\eta)$ be a continuous trapezoidal function of the real variable η that is zero outside the interval $-\xi'_j - \epsilon < \eta < \xi''_j + \epsilon$, that equals one inside the interval $\xi'_j \le \eta \le \xi''_j$, and that is linear on the remaining intervals. (If ξ'_j takes on the improper value $-\infty$, so does $\xi'_j - \epsilon$, etc.)

Let

$$b_\epsilon(\alpha) = \prod_{j=1}^{n} \psi_{j,\epsilon}[X(t_j, \alpha)] \tag{36}$$

and let $b(\alpha)$ be the characteristic function of Q:

$$b(\alpha) = \begin{cases} 1, & \text{if} \quad \alpha \in Q, \\ 0, & \text{otherwise.} \end{cases} \tag{37}$$

Now

$$\lim_{\epsilon \to 0} b_\epsilon(\alpha) = b(\alpha) \tag{38}$$

almost everywhere, and by bounded convergence

$$\lim_{\epsilon \to 0} E\{[b_\epsilon(\alpha) - b(\alpha)]^2\} = 0. \tag{39}$$

As earlier in this section, approximate the random variables $X(t_j, \alpha)$ by series of the form

$$X_{j,m}(\alpha) = \sum_{\mu = 1}^{m} a_{j,\mu} \int_T \phi_\mu(t)\, dX(t, \alpha) \tag{40}$$

with

$$a_{j,\mu} = E\left\{ X(t_j, \alpha) \int_T \phi_\mu(t)\, dX(t, \alpha) \right\} = \int_0^{t_j} \phi_\mu(t)\, dt. \tag{41}$$

As in Equation 27, there exists a subsequence $m_1, m_2, \cdots$, such that

$$\lim_{p \to \infty} X_{j,m_p}(\alpha) = X(t_j, \alpha); \qquad j = 1, \cdots, n \tag{42}$$

for almost all α. (By exercising proper care in defining the subsequence $m_1, m_2, \cdots$, we can use the same subsequence for each value of $j = 1, \cdots, n$.)

By continuity of the trapezoidal functions $\psi_{j,\epsilon}(\eta)$, the function

$$b_{\epsilon,p}(\alpha) = \prod_{j=1}^{n} \psi_{j,\epsilon}[X_{j,m_p}(\alpha)] \tag{43}$$

approaches $b_\epsilon(\alpha)$ almost always as p approaches infinity, and since $|\psi_{j,\epsilon}(\eta)| \le 1$, we have, by bounded convergence,

$$\lim_{p \to \infty} E\{[b_{\epsilon,p}(\alpha) - b_\epsilon(\alpha)]^2\} = 0. \tag{44}$$

By Equation 39, this means that $b(\alpha)$ can be approximated arbitrarily closely in the mean-square sense by functions of the type $b_{\epsilon,p}(\alpha)$. On the other hand, the function $b_{\epsilon,p}$ belongs to the class S with $q = m_p$ and

$$f(u_1, \cdots, u_{m_p}) = \prod_{j=1}^{n} \psi_{j,\epsilon} \left(\sum_{\mu=1}^{m_p} a_{j,\mu} u_\mu \right). \tag{45}$$

This concludes the proof of the lemma.

The lemma suggests that we should be able to build up a complete orthonormal set based on functions of the class S. In the next section we present work of Cameron and Martin which does just this.

3. Fourier-Hermite functions

We denote by $H_n(u)$ the (partially) normalized Hermite polynomial

$$H_n(u) = (-1)^n (n!)^{-\frac{1}{2}} e^{u^2/2} \frac{d^n}{du^n} e^{-u^2/2}, \qquad n = 0, 1, 2, \cdots. \quad (46)$$

As is well known, the set

$$\{(2\pi)^{-\frac{1}{4}} H_n(u) e^{-u^2/4}\} \quad (47)$$

is a complete orthonormal set on $(-\infty, \infty)$,

$$(2\pi)^{-\frac{1}{2}} \int_{-\infty}^{\infty} H_m(u) H_n(u) e^{-u^2/2} \, du = \delta_{mn}; \qquad m, n = 0, 1, 2, \cdots. \quad (48)$$

Using the Hermite polynomials, Cameron and Martin [1947] have defined a Fourier-Hermite set over the space $L_2(0 \le \alpha \le 1)$, as follows.

Definition 1 *Let $\{\phi_\mu(t); \mu = 1, 2, \cdots\}$ be any complete orthonormal set of real functions, each belonging to L_2 over T, and define*

$$\Phi_{m,p}(\alpha) = H_m \left[\int_T \phi_p(t) \, dX(t, \alpha) \right]; \qquad m = 0, 1, 2, \cdots; \quad p = 1, 2, \cdots, \quad (49)$$

and

$$\Psi_{m_1 \cdots m_p}(\alpha) \equiv \Psi_{m_1 \cdots m_p 0 \cdots 0}(\alpha) = \Phi_{m_1, 1}(\alpha) \cdots \Phi_{m_p, p}(\alpha). \quad (50)$$

Cameron and Martin have termed the set of functions 50 the *Fourier-Hermite set*. The index p may be any positive integer, and the subscripts $m_1, \cdots, m_p$ any nonnegative integers; for any particular function Ψ at most a finite number of subscripts may be different from zero. Since the ϕ's belong to $L_2(T)$, the Φ and Ψ are defined for almost all α in $0 \le \alpha \le 1$.

Cameron and Martin have proved the following theorem.

Theorem 2 *The Fourier-Hermite series of any (real or complex) function $F(\alpha)$ of L_2 converges in the mean-square sense to $F(\alpha)$. This means that if $F(\alpha)$ is any function for which*

$$E\{|F(\alpha)|^2\} < \infty, \quad (51)$$

then

$$\lim_{N \to \infty} E\left\{ \left| F(\alpha) - \sum_{m_1, \cdots, m_N = 0}^{N} A_{m_1 \cdots m_N} \Psi_{m_1 \cdots m_N}(\alpha) \right|^2 \right\} = 0, \quad (52)$$

where $A_{m_1 \cdots m_N}$ is the Fourier-Hermite coefficient

$$A_{m_1 \cdots m_N} = E\{F(\alpha) \Psi_{m_1 \cdots m_N}(\alpha)\}. \quad (53)$$

For the proof we first show that the set of functions 50 is orthonormal. For this purpose it is sufficient to show that

$$E\{\Psi_{m_1 \cdots m_n}(\alpha) \Psi_{j_1 \cdots j_n}(\alpha)\} = \delta_{m_1, j_1} \cdots \delta_{m_n, j_n}. \quad (54)$$

In Equation 54 the indices $m_1, \cdots, m_n, j_1, \cdots, j_n$ may be any nonnegative integers. Since zero is an admissible value for any index, there clearly is no loss in taking the same number, n, of m's and j's.

By definitions 49, 50, and Equation 23 of Corollary 1, the left-hand side of Equation 54 is equal to

$$(2\pi)^{-n/2} \int_{-\infty}^{\infty} \cdots \int_{-\infty}^{\infty} H_{m_1}(u_1) H_{j_1}(u_1) e^{-u_1^2/2} \cdots$$
$$H_{m_n}(u_n) H_{j_n}(u_n) e^{-u_n^2/2} du_1 \cdots du_n, \quad (55)$$

and, by Equation 48, expression 55 is equal to the right-hand side of 54. Thus the set 50 is orthonormal. From this, one obtains the Bessel inequality and the best approximation theorem in the usual manner:

$$E\left\{ \left| F(\alpha) - \sum_{m_1, \ldots, m_n = 0}^{n} B_{m_1 \cdots m_n} \Psi_{m_1 \cdots m_n}(\alpha) \right|^2 \right\}$$
$$= E\{|F(\alpha)|^2\} - 2 \operatorname{Re} \sum A_{m_1 \cdots m_n} \bar{B}_{m_1 \cdots m_n} + \sum |B_{m_1 \cdots m_n}|^2$$
$$= E\{|F(\alpha)|^2\} - \sum |A_{m_1 \cdots m_n}|^2 + \sum |A_{m_1 \cdots m_n} - B_{m_1 \cdots m_n}|^2 \quad (56)$$

so that the left-hand member is a minimum whenever $B_{m_1 \cdots m_n} = A_{m_1 \cdots m_n}$ for all $m_1, \cdots, m_n$. Also, since the left member is nonnegative,

$$\sum_{m_1, \cdots, m_n = 0}^{n} |A_{m_1 \cdots m_n}|^2 \leq E\{|F(\alpha)|^2\}. \quad (57)$$

Since the functions of class S defined in the paragraph preceding the statement of Lemma 1 are everywhere dense in $L_2(0, 1)$, it is sufficient for the proof of Theorem 2 to prove it for functions of class S. In fact, let $F(\alpha) \in L_2$, and denote by $A_{m_1 \cdots m_N}$ its Fourier-Hermite coefficients in Equation 53. By Lemma 1, if $\epsilon > 0$, there is a function $f_q(\alpha)$ of class S such that inequality 35 holds. If $A^*_{m_1 \cdots m_N}$ are the Fourier-Hermite coefficients of $f_q(\alpha)$, then assuming that we have Theorem 2 for functions of class S, we can choose N so large that

$$E\left\{\left|f_q(\alpha) - \sum_{m_1, \cdots, m_N = 0}^{N} A^*_{m_1 \cdots m_N} \Psi_{m_1 \cdots m_N}(\alpha)\right|^2\right\} < \epsilon,$$

and by the Minkowski inequality we have

$$E\left\{\left|F(\alpha) - \sum_{m_1, \cdots, m_N = 0}^{N} A^*_{m_1 \cdots m_N} \Psi_{m_1 \cdots m_N}(\alpha)\right|^2\right\} < 4\epsilon.$$

By the best approximation theorem, this same inequality holds with the A^*'s replaced by A's, and hence we obtain Equation 52.

Let then $f_q(\alpha)$ be a function of class S with $f(u_1, \cdots, u_q)$ satisfying condition 34. We first prove the following lemma.

Lemma 2 *For any nonnegative indices $m_1, \cdots, m_p$ we have*

$$E\{f_q(\alpha)\Psi_{m_1 \cdots m_p}(\alpha)\} = \begin{cases} 0, & \text{if} \quad q < p, \quad m_p \neq 0, \\ f_{m_1 \cdots m_q}, & \text{if} \quad p = q, \end{cases} \quad (58)$$

where $f_{m_1 \cdots m_q}$ is the ordinary q-dimensional Hermite coefficient of the L_2 member appearing in condition 34, namely

$$f_{m_1 \cdots m_q} = (2\pi)^{-q/2} \int_{-\infty}^{\infty} \cdots \int_{-\infty}^{\infty} f(u_1, \cdots, u_q) \exp\left[-\frac{(u_1^2 + \cdots + u_q^2)}{4}\right]$$
$$\times \prod_{j=1}^{q} H_{m_j}(u_j) \exp\left[-\frac{u_j^2}{4}\right] du_j. \quad (59)$$

Proof. Since zero is an admissible value for each m_j, we can always arrange in Equation 58 to have at least q subscripts. Thus the two cases $p = q$ and $p > q$ cover all cases. By Equation 23, the left member of Equation 58 is equal to

$$(2\pi)^{-p/2} \int_{-\infty}^{\infty} \cdots \int_{-\infty}^{\infty} f(u_1, \cdots, u_q) \prod_{j=1}^{p} \left[H_{m_j}(u_j) \exp\left[-\frac{u_j^2}{2} \right] du_j \right]. \tag{60}$$

If $p > q$ and $m_p > 0$, the factor $\int_{-\infty}^{\infty} H_{m_p}(u_p) e^{-u_p^2/2} du_p$ comes out and is zero because of the orthogonality relation 48 with $H_0(u_p) \equiv 1$ as the second factor of the integrand. If, on the other hand, $p = q$, then expression 60 agrees with the right member of the statement 59. This yields Lemma 2.

We now return to the proof of Theorem 2. As we have already noted, it is sufficient to prove the theorem for functions of class S. Let $f_q(\alpha)$ be a function of class S, and let $A_{m_1 \cdots m_p}$ be its Fourier-Hermite coefficients,

$$A_{m_1 \cdots m_p} = E\{f_q(\alpha) \Psi_{m_1 \cdots m_p}(\alpha)\}. \tag{61}$$

By definition 59 and Equation 23,

$$E\left\{ \left| f_q(\alpha) - \sum_{m_1, \cdots, m_q = 0}^{N} A_{m_1 \cdots m_q} \Psi_{m_1 \cdots m_q}(\alpha) \right|^2 \right\}$$

$$= (2\pi)^{-q/2} \int_{-\infty}^{\infty} \cdots \int_{-\infty}^{\infty} \left| f(u_1, \cdots, u_q) \exp\left[-\frac{(u_1^2 + \cdots + u_q^2)}{4} \right] \right.$$

$$- \sum_{m_1, \cdots, m_q = 0}^{N} f_{m_1 \cdots m_q} H_{m_1}(u_1) \cdots H_{m_q}(u_q)$$

$$\left. \times \exp\left[-\frac{(u_1^2 + \cdots + u_q^2)}{4} \right] \right|^2 du_1 \cdots du_q \to 0 \qquad \text{as } N \to \infty, \tag{62}$$

due to the completeness of the set $\{\prod_{j=1}^{q} H_{m_j}(u_j) e^{-u_j^2/4}\}$ in $L_2(-\infty, \infty)^q$.

If we now note that $A_{m_1 \cdots m_p}$ defined as in statement 61 is zero when $p > q$ and $m_p > 0$ (see 58), then we see that the number of subscripts on Ψ and A in the left member of Equation 62 may be increased beyond q without changing the sum. Hence we may take N subscripts instead of q, and we have Equation 52 for $F(\alpha) = f_q(\alpha)$, that is, we have Theorem 2 for functions of class S. Hence Theorem 2 holds in general.

Parseval's equality

$$\lim_{N \to \infty} \sum_{m_1, \cdots, m_N = 0}^{N} |A_{m_1 \cdots m_N}|^2 = E\{|F(\alpha)|^2\}$$

follows at once from Theorem 2 and the relation 56 with $B = A$.

4. Distribution functions for certain functionals of the Brownian motion process

In the remaining sections of this chapter we will be concerned with the sample functions $X(t, \alpha)$ of the Brownian motion process, defined for $0 \le t < \infty$ and normalized with $X(0, \alpha) = 0$ for almost all α.

As we have seen in the basic definition,

$$P\{X(t, \alpha) < a\sqrt{t}\} = \frac{1}{\sqrt{2\pi}} \int_{-\infty}^{a} e^{-u^2/2} \, du. \tag{63}$$

It is of interest to know probability distributions of other functionals of the process. Numerous ones have been calculated by various methods. In formula 63 we see (as we already knew) that the distribution of $X(t, \alpha)/t^{1/2}$ is independent of t, $0 < t < \infty$. In the formulas which follow, we shall in general multiply the functional by a suitable power of t to express the results in the same form, that is, independent of t. (In some cases the results were initially obtained for the special value $t = 1$. The result for the general case follows easily from the case $t = 1$ from the fact that for each $\tau > 0$ the function $\tilde{X}(t, \alpha) = X(t\tau, \alpha)/\tau^{1/2}$ is also a sample function of the Brownian motion process.)

For some random variables Z that we shall consider, it is easier to calculate the characteristic function

$$h(\lambda) = E\{e^{i\lambda Z}\}; \qquad -\infty < \lambda < \infty, \tag{64}$$

than the distribution

$$\sigma(a) = P\{Z < a\}; \qquad -\infty < a < \infty. \tag{65}$$

By probability theory, $h(\lambda)$ is expressible in terms of $\sigma(a)$ as follows:

$$h(\lambda) = \int_{-\infty}^{\infty} e^{i\lambda a} \, d\sigma(a). \tag{66}$$

There are various inversion formulas. The one which we shall make use of later states that if $h(\lambda) \in L_1(-\infty, \infty)$, then $\sigma'(a)$ exists and is given by

$$\sigma'(a) = \frac{1}{2\pi} \int_{-\infty}^{\infty} e^{-i\lambda a} h(\lambda) \, d\lambda. \tag{67}$$

(See, for example, Loève [1955, p. 188].)

In the formulas which follow, we shall give the distributions. In discussing certain of them later, however, we shall also list the characteristic functions.

Among the formulas which have been calculated are the following:

I. Denote by $Y(t, \alpha)$ any one of the five random variables

$$|X(t, \alpha)|, \quad M(t, \alpha) = \max_{0 \le \tau \le t} X(\tau, \alpha), \quad -m(t, \alpha) = -\min_{0 \le \tau \le t} X(\tau, \alpha),$$

$$M(t, \alpha) - X(t, \alpha), \quad X(t, \alpha) - m(t, \alpha),$$

and by $\sigma_1(a)$ the distribution function of $Y(t, \alpha)/t^{1/2}$

$$\sigma_1(a) = P\left\{ \frac{Y(t, \alpha)}{t^{1/2}} < a \right\}. \tag{68}$$

We have

$$\sigma_1(a) = 0, \quad \text{for} \quad a \le 0,$$

and

$$\sigma_1(a) = (2/\pi)^{1/2} \int_0^a e^{-u^2/2} \, du, \quad \text{for} \quad a > 0. \tag{69}$$

Next consider the functionals

II.
$$\frac{1}{t^{1/2}} \max_{0 \le \tau \le t} |X(\tau, \alpha)|$$

III.
$$\frac{M(t, \alpha) - m(t, \alpha)}{t^{1/2}},$$

IV.
$$\frac{1}{t} \int_0^t \frac{1 + \operatorname{sgn} X(\tau, \alpha)}{2} \, d\tau,$$

V.
$$\frac{1}{t^{3/2}} \int_0^t |X(\tau, \alpha)| \, d\tau,$$

VI.
$$\frac{1}{t^2} \int_0^t X^2(\tau, \alpha) \, d\tau,$$

and denote by $\sigma_2(a), \cdots, \sigma_6(a)$, respectively, their distribution functions. We have $\sigma_j(a) = 0$ $(a \leq 0); j = 2, \cdots, 6$, and for $a > 0$,

II.
$$\sigma_2(a) = \left(\frac{4}{\pi}\right) \sum_{j=0}^{\infty} \frac{(-1)^j}{2j+1} \exp\left[-\frac{(2j+1)^2 \pi^2}{8a^2}\right],$$

III.
$$\sigma_3(a) = \left[\frac{8}{(2\pi)^{1/2}}\right] \int_0^a \sum_{j=1}^{\infty} (-1)^{j-1} j^2 \exp\left[-\frac{j^2 u^2}{2}\right] du,$$

and

IV.
$$\sigma_4(a) = \begin{cases} (2/\pi) \arcsin a^{1/2}, & 0 < a \leq 1, \\ 1, & 1 < a. \end{cases}$$

V. The real Laplace transform of $\sigma_5(a)$ is given by

$$\int_0^\infty e^{-za} \, d\sigma_5(a) = \sum_{j=1}^{\infty} K_j \exp\left(-\delta_j z^{2/3}\right); \quad z > 0, \quad (70)$$

where δ_j is the jth positive root of the derivative of

$$P(y) = \frac{(2y)^{1/2}}{3}\left[J_{1/3}\left(\frac{2^{1/2} y^{3/2}}{3}\right) + J_{-1/3}\left(\frac{2^{1/2} y^{3/2}}{3}\right)\right], \quad (71)$$

and

$$K_j = \frac{1 + 3 \int_0^{\delta_j} P(y) \, dy}{3\delta_j P(\delta_j)}. \quad (72)$$

VI. $$\sigma_6(a) = \frac{1}{4\pi^{3/2}} \int_0^{a/2} du \cdot u^{-3/2} \int_0^{\pi/2} (\cos v)^{-1/2} \theta_1'\left(\frac{v}{2}, e^{-1/4u}\right) dv,$$

where $\theta_1(z, q)$, $|q| < 1$, is the theta function of the first kind,

$$\theta_1(z, q) = 2 \sum_{n=0}^{\infty} (-1)^n q^{1/4(2n+1)^2} \sin(2n+1)z, \quad \theta_1' \equiv \frac{\partial}{\partial z} \theta_1.$$

As various authors have pointed out, the formulas stated here have interpretations for random walks or for one-dimensional Brownian motion with absorbing barriers. For example, in I, the distribution for $|X(t, \alpha)|$ corresponds to one-dimensional Brownian motion with two symmetrically placed absorbing barriers; if at time t the particle reaches either the point $a\sqrt{t}$ on the right, or the point $-a\sqrt{t}$ on the left, it is absorbed. The distribution gives the probability that the particle remains unabsorbed. The distribution for $M(t, \alpha)$ relates to the motion with an absorbing barrier on the right, that for $m(t, \alpha)$ with an absorbing barrier on the left. For $M(t, \alpha)$, for example, if at any time τ in the time interval $0 \leq \tau \leq t$ the particle reaches the point $a\sqrt{t}$, it is absorbed.

The distribution II corresponds again to two symmetrically placed barriers, the particle being absorbed if at any time τ in $0 \leq \tau \leq t$ it reaches either the point $a\sqrt{t}$ or the point $-a\sqrt{t}$.

Formula III, which gives the distribution of the range $R(t, \alpha) = M(t, \alpha) - m(t, \alpha)$, measures how likely it is that the particle has at most a given spread between how far to the right and how far to the left it goes in the time interval $(0, t)$. That is, it is absorbed if the range at time t exceeds $a\sqrt{t}$.

Formula IV measures the percentage of time the particle remains to the right of the origin, while formulas V and VI refer to absorption when the area under the curve $Y = |X(\tau, \alpha)|$ or $Y = X^2(\tau, \alpha)$, $0 \leq \tau \leq t$, exceeds $at^{3/2}$ or at^2, respectively.

In the next two sections we discuss in more detail the distributions listed above.

5. Limiting distributions, examples of an invariance principle

Derivations of the distributions I of the preceding section are given in Lévy [1948, pp. 209–211]; see also Lévy [1939], Bachelier [1901, 1912, 1913, 1937], Fortet [1943], and Fürth [1917].

Formula II was derived by Lévy [1939] and again by Fortet [1943]; see also Bachelier [1912, 1937], and Fürth [1917].

Formula III was derived by Feller [1951], IV by Erdös and Kac [1947], V by Kac [1946], and VI by Cameron and Martin [1944].

Related to formula I_2 (i.e., the part giving the distribution of $M(t, \alpha)$), and formulas II, IV, V, and VI, Erdös and Kac [1946, 1947] have proved the following five limit theorems. Let $X_1, X_2, \cdots$, be

independent identically distributed random variables each having
mean zero and variance one. We denote by S_n the nth cumulative
sum,

$$S_n = X_1 + \cdots + X_n. \tag{73}$$

It follows that

$$\lim_{n \to \infty} P\{\max (S_1, \cdots, S_n) < an^{1/2}\} = \sigma_1(a) \tag{74}$$

$$\lim_{n \to \infty} P\{\max (|S_1|, \cdots, |S_n|) < an^{1/2}\} = \sigma_2(a) \tag{75}$$

$$\lim_{n \to \infty} P\{N_n < an\} = \sigma_4(a) \tag{76}$$

(N_n denotes the number of S_j's, $1 \le j \le n$, which are positive),

$$\lim_{n \to \infty} P\{|S_1| + \cdots + |S_n| < an^{3/2}\} = \sigma_5(a) \tag{77}$$

and

$$\lim_{n \to \infty} P\{S_1^2 + \cdots + S_n^2 < an^2\} = \sigma_6(a). \tag{78}$$

Under the same hypotheses, Feller [1951] showed that

$$\lim_{n \to \infty} P\{M_n - m_n < an^{1/2}\} = \sigma_3(a), \tag{79}$$

where

$$M_n = \max [0, S_1, \cdots, S_n], \qquad m_n = \min [0, S_1, \cdots, S_n]. \tag{80}$$

The relation 77 was also derived by Kac [1946].

The condition that the X's are identically distributed can be
replaced by a weaker one, namely by the assumption that the X's
are such that the central limit theorem is applicable. Erdös and Kac
[1947] actually proved statement 76 for this more general case, and
they wrote in a footnote in their 1946 paper that this could be done
for statements 74, 75, 77, and 78.

In the result 74, if one temporarily fixes n and chooses X_j to be

$$(n/t)^{1/2}[X(jt/n, \alpha) - X((j-1)t/n, \alpha)]; \qquad j = 1, \cdots, n, \tag{81}$$

then

$$S_j = (n/t)^{1/2} X(jt/n, \alpha); \qquad j = 1, \cdots, n, \tag{82}$$

and 74 states that

$$\lim_{n \to \infty} P\{\max [X(t/n, \alpha), X(2t/n, \alpha), \cdots, X(nt/n, \alpha)] < at^{1/2}\} = \sigma_1(a), \tag{83}$$

which implies formula I_2. Similarly, statement 75 implies formula II.

For statements 76 through 78, if one again chooses X_j as we did in expression 81 so that the S_j have the value of Equation 82, then

$$\lim_{n \to \infty} \frac{N_n}{n} = \frac{1}{t} \int_0^t \frac{1 + \text{sgn } X(\tau, \alpha)}{2} \, d\tau,$$

$$\lim_{n \to \infty} \left[\sum_{j=1}^n \frac{|S_j|}{n^{3/2}} \right] = \lim_{n \to \infty} \frac{1}{t^{3/2}} \frac{t}{n} \sum_{j=1}^n \left| X\left(\frac{jt}{n}, \alpha\right) \right|$$

$$= \frac{1}{t^{3/2}} \int_0^t |X(\tau, \alpha)| \, d\tau,$$

and

$$\lim_{n \to \infty} \left[\sum_{j=1}^n \frac{S_j^2}{n^2} \right] = \lim_{n \to \infty} \frac{1}{t^2} \frac{t}{n} \sum_{j=1}^n X^2\left(\frac{jt}{n}, \alpha\right)$$

$$= \frac{1}{t^2} \int_0^t X^2(\tau, \alpha) \, d\tau.$$

Thus statements 76 through 78 imply formulas IV, V, and VI, respectively.

As Erdös and Kac pointed out, the result 76 was proved in the binomial case $P\{X_j = 1\} = P\{X_j = -1\} = \frac{1}{2}$ by Lévy [1939], who also indicated that it is true for more general random variables.

The proof of statement 78 given by Erdös and Kac made use of the formula VI derived by Cameron and Martin [1944].

We shall close this section with a few comments on Feller's formulas III and 79. In the next section we discuss in more detail formula VI and the related characteristic function, as well as a similar formula for the random variable $\int_0^1 p(t)X^2(t, \alpha) \, dt$, where $p(t)$ is a given function, positive and continuous on [0, 1]. These results will be related to certain differential and integral equations.

We return now to a discussion of Feller's formula III and the related formula 79. Lévy [1948, p. 213] showed for $a < 0$ and $b > 0$ that

$$P\{at^{1/2} < m(t, \alpha), M(t, \alpha) < bt^{1/2}, x \le X(t, \alpha) < x + dx\}$$

$$= \left[\frac{dx}{(2\pi)^{1/2}} \right] \sum_{n=-\infty}^{\infty} \left\{ \exp\left[-\frac{(x - 2n\ell)^2}{2} \right] - \exp\left[-\frac{(x - 2a + 2n\ell)^2}{2} \right] \right\},$$

$$(84)$$

where $\ell = b - a$. This formula was also obtained by Feller [1951, p. 430] who used it to derive formula III. Feller's work on formula III was related in part to his interest in the distribution of the range $R_n = M_n - m_n$, where M_n and m_n are as in definitions 80 and 73, with the X_j's independent identically distributed random variables, each with mean zero and variance one. Feller points out that since the sums S_n are asymptotically normally distributed, the asymptotic distribution of the range is independent of the underlying distribution of the X_j's. He therefore considered the case in which the random variables are normal, and in this case the sum S_n can be considered as the value at time $t = n$ of the sample function $X(t, \alpha)$ of the Brownian motion process. Since the sequence S_j is a subset of the values assumed by $X(t, \alpha)$, the range R_n is not larger than the range at time $t = n$ of $X(t, \alpha)$:

$$R_n \leq R(n, \alpha) \equiv M(n, \alpha) - m(n, \alpha),$$

and as Feller states, for large n the two ranges R_n and $R(n, \alpha)$ will be practically the same. He then derives formula III which gives the exact distribution of the range $R(t, \alpha) \equiv M(t, \alpha) - m(t, \alpha)$ of the continuous variable $X(t, \alpha)$. Feller uses III to calculate precisely the mean and variance of $R(n, \alpha)$, namely

$$E\{R(n, \alpha)n^{-1/2}\} = 2(2/\pi)^{1/2};$$
$$\mathrm{Var}\,\{R(n, \alpha)n^{-1/2}\} = 4\,(\log 2 - 2/\pi).$$

(In the same paper Feller also considers the adjusted range R_n^* of the random variables $S_j^* = S_j - jS_n/n$; $j = 1, \cdots, n$, and the range $R^*(t, \alpha)$ of the variable $X^*(t, \alpha) = X(t, \alpha) - tX(T, \alpha)/T$, $0 < t < T$, and he obtains the same types of results for these ranges. As he points out, the latter set of results has an advantage in certain applications.)

Darling and Siegert [1953] have developed a formula for the range

$$R(Z, t) = \sup_{0 \leq \tau \leq t} Z(\tau) - \inf_{0 \leq \tau \leq t} Z(\tau),$$

where $Z(t)$ is a stochastic process which satisfies very general conditions.

6. Other distributions and other examples of the invariance principle

The method used by Erdös and Kac in proving the limit theorems described in the preceding section consisted of first proving that the limiting distribution in each case is independent of the underlying distribution of the sequence $X_1, X_2, \cdots$, and then calculating the limiting distribution by making a convenient choice of the underlying distribution. The first step in the proof of such a limit theorem is now often referred to as the Erdös-Kac invariance principle.

Fortet [1949] and Mark [1949] obtained further results in which the invariance principle holds, and Kac [1949] derived a general theorem of this nature. Donsker [1951] proved a very general invariance principle which included previous results in this direction. Before describing these results we shall discuss briefly formula VI of Section 4 and a related formula for the characteristic function of the random variable

$$\int_0^1 p(t)X^2(t, \alpha)\, dt. \tag{85}$$

This discussion will be related to the invariance principle which follows.

As was remarked earlier, formula VI was obtained by Cameron and Martin [1944]. We shall now describe their derivation of this formula. Their proof depended upon a formula on transformations of Wiener integrals under linear transformations

$$T: Y(t, \alpha) = X(t, \alpha) + \int_0^1 K(t, s)X(s, \alpha)\, ds = X(t, T\alpha) \tag{86}$$

in which the kernel satisfies suitable conditions. (A precise statement of this result, together with the proof, is given in the paper by Cameron and Martin [1945a].)

If $F(\alpha)$ is any measurable function (for example) $F(\alpha) = \int_0^1 X^2(t, \alpha)\, dt$ or $F(\alpha) = \max_{0 \le t \le 1} |X(t, \alpha)|$, etc., then under the transformation equation 86, one has

$$E\{F(\alpha)\} = |D|E\{F(T\alpha)\, e^{-\Phi(\alpha)}\}, \tag{87}$$

where

$$D = 1 + \sum_{\mu=1}^{\infty} \frac{1}{\mu!} \int_0^1 \cdots \int_0^1 \begin{vmatrix} K(s_1, s_1) \cdots K(s_1, s_\mu) \\ \cdot \quad \cdot \quad \cdot \quad \cdot \quad \cdot \quad \cdot \\ K(s_\mu, s_1) \cdots K(s_\mu, s_\mu) \end{vmatrix} ds_1 \cdots ds_\mu, \quad (88)$$

and Φ is suitably defined. For example, if we take the special case of T

$$T: \quad Y(t, \alpha) = X(t, \alpha) + \lambda \int_0^1 [\tan \lambda(s - 1)] \cdot X(s, \alpha) \, ds, \quad (89)$$

then

$$|D| = [\cos 2^{\frac{1}{2}}\lambda]^{\frac{1}{2}}, \qquad \Phi(\alpha) = -\lambda^2 \int_0^1 X^2(t, \alpha) \, dt, \quad (90)$$

for $-\pi/2 < 2^{\frac{1}{2}}\lambda < \pi/2$. If now we take $F(\alpha) \equiv 1$, then the transformation formula 87 yields

$$1 = [\cos 2^{\frac{1}{2}}\lambda]^{\frac{1}{2}} E\left\{ \exp\left[\lambda^2 \int_0^1 X^2(t, \alpha) \, dt \right] \right\}, \quad -\pi 2^{\frac{1}{2}}/4 < \lambda < \pi 2^{\frac{1}{2}}/4.$$

$$(91)$$

By analytic continuation Equation 91 holds for λ complex with $\mathrm{Re}\,(\lambda^2) < \pi^2/8$. Thus

$$E\left\{ \exp\left[\lambda \int_0^1 X^2(t, \alpha) \, dt \right] \right\} = [\sec (2\lambda)^{\frac{1}{2}}]^{\frac{1}{2}}, \quad \mathrm{Re}\,\lambda < \pi^2/8. \quad (92)$$

This gives the characteristic function of the random variable $\int_0^1 X^2(t, \alpha) \, dt$, namely

$$E\left\{ \exp\left[i\mu \int_0^1 X^2(t, \alpha) \, dt \right] \right\} = [\sec (2i\mu)^{\frac{1}{2}}]^{\frac{1}{2}}, \qquad -\infty < \mu < \infty,$$

$$(93)$$

where the right-hand side has the determination which is positive when μ is purely imaginary and contained in the interval $(-i\pi^2/8, 0)$. (The characteristic function obtained by Cameron and Martin [1945a] was actually $[\sec (i\mu)^{\frac{1}{2}}]^{\frac{1}{2}}$ which corresponds to the different normalization which they used, namely $E\{X^2(t, \alpha)\} = t/2$ rather than $E\{X^2(t, \alpha)\} = t$.)

The characteristic function 93 can also be evaluated directly by a modification of the method of Paley and Wiener [1934], using the statistical independence of the random variables

$$\int_0^1 \phi_\mu(t)\, dX(t, \alpha); \qquad \mu = 1, 2, \cdots$$

for $\{\phi_\mu(t)\}$ an orthonormal set of real functions on $(0, 1)$, each belonging to $L_2(0, 1)$. It has also been obtained by various other authors, for example by Bochner [1947], Fortet [1949], and Lévy [1953].

It is very easy to show that the characteristic function 93 belongs to $L_1(-\infty, \infty)$ (see for example Cameron and Martin [1944]), and hence formula 67 enables one to obtain the distribution function $\sigma_6(a)$ for the random variable $\int_0^1 X^2(t, \alpha)\, dt$. Clearly $\sigma_6(a) = 0$ for $a \le 0$, and by Equation 67 we have for $a > 0$

$$\sigma_6(a) = \frac{1}{2\pi} \int_0^a d\beta \int_{-\infty}^{\infty} [\sec (2i\mu)^{\frac{1}{2}}]\, e^{-i\mu\beta}\, d\mu$$

$$= \frac{1}{2\pi} \int_0^{a/2} d\gamma \int_{-\infty}^{\infty} [\sec (-i\gamma)^{\frac{1}{2}}]^{\frac{1}{2}}\, e^{i\lambda\gamma}\, d\lambda. \tag{94}$$

Cameron and Martin [1944, pp. 205–207] have shown that

$$\int_{-\infty}^{\infty} [\sec (-i\gamma)^{\frac{1}{2}}]^{\frac{1}{2}}\, e^{i\lambda\gamma}\, d\lambda = \frac{1}{2\pi^{\frac{1}{2}}\gamma^{\frac{3}{2}}} \int_0^{\pi/2} \frac{\theta_1'\left(\frac{s}{2}, e^{-\frac{1}{4}\gamma}\right)}{\sqrt{\cos s}}\, ds, \qquad \gamma > 0. \tag{95}$$

This, with Equation 93, yields formula VI of Section 4 for $t = 1$. The result for arbitrary positive values of t follows as indicated at the beginning of Section 4.

As a generalization of formula VI Cameron and Martin [1945b] proved the following theorem: *Let $p(t)$ be continuous and positive on $0 \le t \le 1$, and let λ_0 be the least characteristic value of the differential equation*

$$f''(t) + \lambda p(t)f(t) = 0 \tag{96}$$

subject to the boundary conditions

$$f(0) = f'(1) = 0. \tag{97}$$

If $-\infty < 2\mu < \lambda_0$, *and* $f_{2\mu}(t)$ *is any nontrivial solution of* 96 *satisfying* $f_{2\mu}'(1) = 0$, *we have*

$$E\left\{\exp\left[\mu \int_0^1 p(t)X^2(t, \alpha)\, dt\right]\right\} = [f_{2\mu}(1)/f_{2\mu}(0)]^{1/2}. \qquad (98)$$

The proof follows from the general linear transformation theorem cited earlier in this section by specializing 86 to be

$$Y(t, \alpha) = X(t, \alpha) - \int_0^t \frac{f_{2\mu}'(s)}{f_{2\mu}(s)} X(s, \alpha)\, ds, \qquad (99)$$

where μ is fixed, $-\infty < 2\mu < \lambda_0$, and $f_{2\mu}(t)$ is any nontrivial solution of Equation 96 satisfying $f_{2\mu}'(1) = 0$. The formula 98 has been derived by simpler methods by other mathematicians. For example Kac showed that

$$E\left\{\exp\left[iv \int_0^1 p(t)X^2(t, \alpha)\, dt\right]\right\} = \prod_{j=1}^{\infty} (1 - 2iv\lambda_j)^{-1/2} \qquad (100)$$

where $\lambda_1, \lambda_2, \lambda_3, \cdots$, are the characteristic values of the integral equations

$$\int_0^1 \min(s, t)p(t)f(t)\, dt = \lambda f(s) \qquad (101)$$

or, what is the same, of the differential equation 96 subject to the boundary conditions 97.

Kac and Siegert [1947] proved the following invariance theorem for this case, namely if $X_1, X_2, \cdots$, are independent identically distributed random variables, each with mean zero and variance one, and if S_j is the jth cumulative sum in definition 73, then

$$\lim_{n \to \infty} P\left\{\frac{1}{n^2} \sum_{j=1}^{n} p(j/n)S_j^2 < a\right\} = \sigma_7(a), \qquad (102)$$

where the characteristic function of $\sigma_7(a)$ is given by Equation 100.

Under the same conditions on the X_j's, Mark [1949] derived a formula similar to statement 102 when the condition that $p(t)$ be positive on $[0, 1]$ is replaced by the condition that it have at most a denumerable number of zeros in the interval. In this case he showed that the characteristic function of the distribution of statement 102 is $[D(2iv)]^{-1/2}$, where $D(\mu)$ is the Fredholm determinant associated with the integral equation

$$f(s) - \mu \int_0^1 \min\,(s,t)p(t)f(t)\,dt = 0. \tag{103}$$

Mark also proved several other limit theorems (invariance principles) which we will not cite here.

Another example of the invariance principle was derived by Fortet [1949]. With the X_j's and S_j's defined as above, denote by I_m^n and J_m^n the random variables

$$I_m^n = n^{-m/2-1} \sum_{j=1}^n S_j^m, \qquad J_m^n = n^{-m/2-1} \sum_{j=1}^n |S_j|^m. \tag{104}$$

Under the hypothesis that the moment $E\{X_j^{m^*}\}$ exists, where m^* is the smallest even integer such that $m^* \geq m$, Fortet proved that

$$\lim_{n \to \infty} P\{I_m^n < a\} = P\left\{\int_0^1 X^m(t, \alpha)\,dt < a\right\} \tag{105}$$

and

$$\lim_{n \to \infty} P\{J_m^n < a\} = P\left\{\int_0^1 |X(t, \alpha)|^m\,dt < a\right\}. \tag{106}$$

For $m = 1$ and $m = 2$, the assumption on the existence of the m^*th moment adds nothing, and statement 106 for $m = 1$ and statement 105 for $m = 2$ are equivalent to statements 77 and 78, respectively. Fortet also gives (pp. 184–186) an alternative proof of the result 93 and, as he states, his method permits him to prove the Erdös-Kac invariance principle, statements 74 and 75. (In the next section the condition on the higher moment, of order m^*, will be removed.)

We close this section with a brief mention of another important but different class of asymptotic theorems, proved by Kallianpur and Robbins [1953] and generalized by Derman [1954]. Let $f(x)$ and $g(x)$ be two bounded, integrable, real-valued functions on $-\infty < x < \infty$, and set

$$\bar{f} = \int_{-\infty}^\infty f(x)\,dx, \qquad \bar{g} = \int_{-\infty}^\infty g(x)\,dx. \tag{107}$$

Kallianpur and Robbins have proved the following two theorems:

1°. *If $\hat{f} \neq 0$, then for every a*

$$\lim_{T \to \infty} P\left\{\frac{1}{\hat{f}T^{\frac{1}{2}}} \int_0^T f[X(t, \alpha)] \, dt \leq a\right\} = \sigma_1(a), \qquad (108)$$

where

$$\sigma_1(a) = \begin{cases} 0, & \text{if} \quad a \leq 0, \\ (2/\pi)^{\frac{1}{2}} \int_0^a e^{-v^2/2} \, dv, & \text{if} \quad a > 0. \end{cases} \qquad (109)$$

2°. *If $\bar{g} \neq 0$, then*

$$\lim_{T \to \infty} \frac{\int_0^T f[X(t, \alpha)] \, dt}{\int_0^T g[X(t, \alpha)] \, dt} = \frac{\hat{f}}{\bar{g}} \qquad (110)$$

in probability.

Derman's generalization of 2° is the following:

3°. *If f and g are any two Baire functions, with finite integrals given in definition 107 and if $\bar{g} \neq 0$, then the result 110 holds with probability one.*

7. More general formulations of the invariance principle

As earlier, let $X_1, X_2, \cdots,$ be independent, identically distributed random variables, each with mean zero and variance one, and denote by S_j the jth cumulative sum of definition 73. Let $V(x)$ be a nonnegative function on $(-\infty, \infty)$ which is continuous (or more generally piecewise continuous with a finite number of discontinuities.) Kac [1949] considered the two distributions,

$$P\left\{\frac{t}{n} \sum_{j=1}^n V\left[S_j \cdot \left(\frac{t}{n}\right)^{\frac{1}{2}}\right] < a\right\} = \sigma_n(a, t) \qquad (111)$$

and

$$P\left\{\int_0^t V[X(\tau, \alpha)] \, d\tau < a\right\} = \sigma(a, t). \qquad (112)$$

If the X_j are chosen as in 81 so that the S_j have the value of definition 82, then

$$\frac{t}{n} \sum_{j=1}^{n} V\left[S_j \cdot \left(\frac{t}{n}\right)^{1/2}\right] = \frac{t}{n} \sum_{j=1}^{n} V\left[X\left(\frac{jt}{n}, \alpha\right)\right]$$

$$\rightarrow \int_{0}^{t} V[X(\tau, \alpha)] \, d\tau \qquad \text{as} \qquad n \rightarrow \infty \qquad (113)$$

and thus for this case

$$\lim_{n \to \infty} \sigma_n(a, t) = \sigma(a, t) \qquad (114)$$

at each point of continuity of $\sigma(a, t)$.

The results cited in Section 5 furnish several examples in which statement 114 holds for the X's defined as at the beginning of this section.

Under the conditions stated above on $V(x)$ and under the further condition that the common distribution of the X_j's is normal, Kac [1949] proved that statement 114 holds. In the same paper he indicated that possibly under further restrictions on the X's, relation 114 would hold regardless of the common distribution of the X's.

Donsker [1951] proved a very general version of the invariance principle which contained the various results previously obtained in this direction. As Kac [1951] pointed out, Donsker's general result enables one to give the best conditions on $V(x)$ which make statement 114 true without further restrictions on the X_j's, other than that they be independent, identically distributed random variables each with mean zero and variance one. Under these conditions on the X's, Donsker's theorem may be summarized as follows.

Let R be the space of real-valued functions $g(t)$ which are continuous except possibly for finite number of finite jumps on $[0, 1]$, and let $F(g)$ be defined on R and continuous (in the uniform topology) at almost all points of C, where C is the space of all continuous functions $x(t)$ defined on $0 \leq t \leq 1$ and vanishing at the origin, with the Wiener measure imposed on C. Finally, denote by $b_n(t)$ the random step function

$$b_n(t) = n^{-1/2} S_j \qquad \text{if} \qquad (j-1)/n \leq t \leq j/n; \qquad j = 1, \cdots, n. \qquad (115)$$

It then follows that

$$\lim_{n \to \infty} P\{F(b_n) < a\} = P\{F[X(\cdot, \alpha)] < a\} \qquad (116)$$

at every point of continuity of the distribution function of $F[X(\cdot, \alpha)]$.

This theorem not only implies the previous results on the invariance principle but also strengthens some of them. For example, it yields Fortet's results, statements 105 and 106, without the assumption of the higher moments beyond the first two.

Billingsley [1956] has extended to the dependent case Donker's generalized form of the Erdös-Kac invariance principle. We will not have occasion in this treatment to make use of Billingsley's interesting extension.

Having described the invariance principle related to the distribution 111, we return to a discussion of Kac's work. Kac [1949] (see also Kac [1951, pp. 190–196] and [1959, pp. 165–175]) has proved the following result.

Theorem 3 *Let* $V(x)$ *be nonnegative and continuous on* $(-\infty, \infty)$. *Then for each* $t > 0$

$$E\left\{ \exp\left[-\int_0^t V[X(\tau, \alpha)] \, d\tau \right] \right\} = \int_{-\infty}^{\infty} Q(x, t) \, dx, \qquad (117)$$

where $Q(x, t)$ *satisfies the integral equation*

$$Q(x, t) + (2\pi)^{-\frac{1}{2}} \int_0^t \int_{-\infty}^{\infty} (t - \tau)^{-\frac{1}{2}} \exp\left[\frac{-\frac{1}{2}(x - \xi)^2}{t - \tau} \right]$$
$$\cdot V(\xi) Q(\xi, \tau) \, d\xi \, d\tau = (2\pi t)^{-\frac{1}{2}} \exp\left[\frac{-\frac{1}{2}x^2}{t} \right]. \qquad (118)$$

Kac first proves this under the additional assumption that V is bounded, $0 \leq V(x) < M$. He then shows that if one integrates over the part of the Wiener space C defined by the condition

$$a < X(t, \alpha) < b, \qquad (119)$$

the relation

$$E\left\{ \exp\left[-\int_0^t V[X(\tau, \alpha)] \, d\tau \right]; a < X(t, \alpha) < b \right\} = \int_a^b Q(x, t) \, dx,$$
$$(120)$$

holds. This implies in particular that $Q(x, t)$ is nonnegative and this enables him to remove the boundedness restriction by the complete additivity of the Wiener integral. The integral equation 118 implies that Q satisfies the differential equation

$$\frac{\partial Q}{\partial t} = \frac{1}{2} \frac{\partial^2 Q}{\partial x^2} - V(x)Q, \qquad (121)$$

and the relation 120 implies the initial condition

$$\lim_{t \to 0} \int_{-\epsilon}^{\epsilon} Q(x, t) \, dx = 1. \qquad (122)$$

If one forms the Laplace transform

$$\psi(x) \equiv \psi(x, s) = \int_{-\infty}^{\infty} Q(x, t) \exp(-st) \, dt, \qquad s > 0, \qquad (123)$$

then Kac shows that ψ satisfies the differential equation

$$\tfrac{1}{2}\psi''(x) - [s + V(x)]\psi(x) = 0 \qquad (124)$$

subject to the conditions

(a) $\psi \to 0$ as $x \to \pm\infty$,
(b) ψ' is continuous except at $x = 0$,
(c) $\psi'(-0) - \psi'(+0) = 2$.

As Kac remarks, only minor modifications are necessary if $V(x)$ is allowed to have a finite number of discontinuities.

Thus one has an evaluation of the Wiener integral (or expected value) on the left-hand side of Equation 117 (or 120) in terms of the solution of the Schrödinger type equation 124 with the conditions (a), (b), and (c).

Kac [1949, pp. 11–13] illustrated his general theory by two examples, namely $V(x) = x^2$ and $V(x) = (1 + \text{sgn } x)/2$, and thus obtained proofs by new methods of the formulas for these two cases.

We return to the general case as stated in Theorem 3. If one adds the assumptions that

$$V(x) \to \infty \qquad \text{as} \qquad x \to \pm\infty, \qquad (125)$$

then, as Kac points out, the eigenvalue problem

$$\tfrac{1}{2}\psi'' - V(x)\psi = -\lambda\psi, \qquad \psi \in L_2(-\infty, \infty) \qquad (126)$$

has a discrete spectrum

$$\lambda_1, \lambda_2, \cdots, \qquad (127)$$

with corresponding normalized eigenfunctions

$$\psi_1(x), \psi_2(x), \cdots, \qquad (128)$$

and the function Q can be written in the form

$$Q(x, t) = \sum_{j=1}^{\infty} [\exp(-\lambda_j t)]\psi_j(x)\psi_j(0). \qquad (129)$$

Thus the integral over function space, or expected value, given on the left-hand side of 117 is expressed in terms of the eigenvalues and eigenfunctions of the Schrödinger eigenvalue problem 126. In the three references cited earlier Kac [1949, 1951, 1959] draws numerous interesting conclusions from these results.

4

Matrix Factorization and Prediction*

1. [One-dimensional motivations]

This note will deal primarily with binary matrices whose elements are functions of a variable θ which is to run between $(-\pi, \pi)$. It represents an extension of certain well-known theorems due to Szegö[1] and the author, concerning scalar functions of θ. The fundamental theorem is the following:

[Theorem 1]   *Let $F(\theta)$ be nonnegative and belong to Lebesgue class L over $(-\pi, \pi)$. Then a necessary and sufficient condition for us to be able to write*

$$F(\theta) = |\phi(\theta)|^2, \tag{1}$$

where

$$\phi(\theta) = \sum_0^\infty a_n \, e^{in\theta} \tag{2}$$

and

$$\sum_0^\infty |a_n|^2 < \infty, \tag{3}$$

* The first four sections of this chapter comprise a reprinting of the first four sections of the article, "On the Factorization of Matrices," by Norbert Wiener, that appeared in *Commentarii Mathematici Helvetici* [1955]. Footnotes have been supplied by the editor. Minor deviations in format from the original article have been made for the sake of clarity, but all such changes have been indicated by square brackets. Minor changes that can be accounted for by proofreading errors have also been made and are indicated in the text by square brackets, or in a display formula by an asterisk following the accompanying formula number. The original article was dedicated as follows: *To Professor Plancherel, the founder of the precise theory of the Fourier integral and the inspirer of my work on harmonic analysis.*

[1] See G. Szegö [1939, Chapter 10].

is that

$$\int_{-\pi}^{\pi} |\log F(\theta)| \, d\theta \tag{4}$$

be finite. It is then possible to choose the coefficients a_n in such a manner that

$$\sum a_n z^n \tag{5}$$

has no zeros inside the unit circle.

Let α be an arbitrary real number between 0 and 1. Let it be represented in the binary scale by the expression:

$$\alpha = .\alpha_1 \, \alpha_2 \, \alpha_3 \cdots . \tag{6}$$

Let these digits be renumbered:

$$.\beta_0 \, \beta_1 \, \beta_{-1} \, \beta_2 \, \beta_{-2} \cdots$$

and so on. Let

$$B_n(\alpha) = 2\beta_n - 1. \tag{7}$$

It will follow that the transformation of α which changes $B_n(\alpha)$ into $B_{n+1}(\alpha)$ for all values of α lying between 0 and 1, and all values of n, is a measure-preserving transformation T. We may write

$$B_{n+1}(\alpha) = B_n(T\alpha). \tag{8}$$

This transformation[2] T is not indeed well defined for all values of α but is well defined for all values of α with the exception of a set of measure 0.

If we start with any function $\phi(\theta)$ belonging to L_2 and containing no negative frequencies, we can represent it, as I have said before, by the sequence of coefficients a_n where:

$$\sum_0^\infty |a_n|^2 < \infty. \tag{9}$$

---

[2] For the use of measure-preserving transformations as a constructional device in probability theory, see Chapter 1. For an earlier reference to the use of such transformations, see Paley and Wiener [1934, Chapter 9]. For questions concerning the generality of a stationary process such as that which follows, in the form $\{\cdots f(T^1\alpha), f(\alpha), f(T^{-1}\alpha)\cdots\}$, see Doob [1953, pp. 452–457].

Under these circumstances, it can be proved that

$$\sum_0^\infty a_n B_{-n}(\alpha) \tag{10}$$

will converge in the mean to a function of α which we shall call $f(\alpha)$. The function $f(\alpha)$ will then belong to L_2 over the interval $(0, 1)$. If we consider the projection of any function $g(\alpha)$ belonging to L_2 on the closure of the set of

$$f(T^{-n}\alpha), \quad f(T^{-n-1}\alpha), \quad f(T^{-n-2}\alpha), \cdots, \tag{11}$$

this will converge in the mean to 0. It will obviously be the same as the projection of g on the closure of the set of functions $B_{-n}(\alpha)$, $B_{-n-1}(\alpha), \cdots$. That is, it will be the function

$$\sum_{\nu=0}^\infty B_{-\nu}(\alpha) \int_0^1 g(\beta) B_\nu(\beta) \, d\beta, \tag{12}$$

and will have as the integral of the square of its absolute value

$$\sum_{\nu=0}^\infty \left| \int_0^1 g(\beta) B_{-\nu}(\beta) \, d\beta \right|^2. \tag{13}$$

This leads us immediately to the closely related

[Theorem[3] 2] *Let us assume in general that $f(\alpha)$ is any function whatever of the variable α which lies on $(0, 1)$. Let T be any measure-preserving transformation of α into itself. Let the projection of $f(\alpha)$ on the set of functions*

$$f(T^{-n}\alpha), \, f(T^{-n-1}\alpha), \cdots, \tag{14}$$

converge in the mean to 0 as n becomes infinite. Then there exists a function $h(\alpha)$ which is normalized which is linearly dependent on the set of functions

$$f(\alpha), \, f(T^{-1}\alpha), \cdots,$$

---

[3] The function $h(\alpha)$ of this theorem, when it appears in prediction theory, is called the *innovation function*, a term generally attributed to Wiener. The expansion of $f(\alpha)$ in terms of $h(T^{-n}\alpha)$, as stated after the theorem, is at the basis of single series prediction. See Section 5. The phrase, "Projection of $f(\alpha)$ on the set of functions," is an abbreviation for the longer phrase, "Projection of $f(\alpha)$ on the closed linear manifold generated by the set of functions."

and which is orthogonal to all functions

$$f(T^{-1}\alpha),\ f(T^{-2}\alpha),\ \cdots. \tag{15}$$

It will follow that the functions $h(T^n\alpha)$ are a normal and orthogonal set, and it can be proved that $f(\alpha)$ will be equal to

$$f(\alpha) = \sum_0^\infty h(T^{-n}\alpha) \int_0^1 f(\beta)\overline{h(T^{-n}\beta)}\ d\beta \tag{16}*$$

as a limit in the mean. The function

$$\sum_0^\infty z^n \int_0^1 f(\beta)\overline{h(T^{-n}\beta)}\ d\beta \tag{17}$$

will be analytic inside the unit circle and will have no zeros there. Taken around any circle concentric with the unit circle but of smaller radius, the integral of the absolute square of this function will be uniformly bounded.

The statement in the hypothesis that $f(\alpha)$ is asymptotically orthogonal to the closure of

$$f(T^{-n}\alpha),\ f(T^{-n-1}\alpha),\ \cdots,$$

as n becomes infinite is obviously a statement which merely concerns the autocorrelation coefficients

$$\int_0^1 f(T^n\alpha)\overline{f(\alpha)}\ d\alpha. \tag{18}$$

If then, these are of the form [4]

$$\frac{1}{2\pi} \int_{-\pi}^\pi F(\theta)\ e^{-in\theta}\ d\theta, \tag{19}$$

---

* Asterisk ff. denotes proofreading corrections to original article.

[4] The function $F(\theta)$ is called the *spectral density* of the process $\{f(T^n\alpha),\ n = \cdots -1, 0, 1, \cdots\}$. Under more general conditions, we can write:

$$\int_0^1 f(T^n\alpha)\overline{f(\alpha)}\ d\alpha = \frac{1}{2\pi} \int_{-\pi}^\pi e^{-in\theta}\ dG(\theta),$$

where $G(\theta)$ is the *spectral distribution* of the process. Thus, the existence of the spectral density $F(\theta)$ (as an L_2-function with Fourier coefficients $\int_0^1 f(T^n\alpha)f(\alpha)\ d\alpha$) is equivalent to the absolute continuity of the spectral distribution function $G(\theta)$.

we can reduce this case to the particular case in which we have derived $f(\alpha)$ from $\phi(\theta)$ by means of the B's.

2. [Two-dimensional conditions]

Now, let us start with two functions of class L_2, $f_1(\alpha)$, $f_2(\alpha)$. Parenthetically, let me remark that these both are to belong to L_2 and that we have one single transformation T of α into itself which preserves measure. Let the remote pasts of both f_1 and f_2 be asymptotically orthogonal to f_1 and f_2, which will be the case if[5] $F_1(\theta)$ and $F_2(\theta)$ are respectively the functions belonging to L_2 with Fourier coefficients

$$\int_0^1 f_1(T^n\alpha)f_1(\alpha)\,d\alpha \tag{20}$$

and

$$\int_0^1 f_2(T^n\alpha)f_2(\alpha)\,d\alpha, \tag{21}$$

and [   ]

$$\int_{-\pi}^{\pi} |\log F_1(\theta)|\,d\theta < \infty, \qquad \int_{-\pi}^{\pi} |\log F_2(\theta)|\,d\theta < \infty. \tag{22}$$

Under these circumstances we shall have two normalized functions $h_1(\alpha)$ and $h_2(\alpha)$ such that h_1 is linearly dependent on f_1 and $f_1(T^{-n}\alpha)$ and orthogonal to all functions $f_1(T^{-n}\alpha)$ where n is positive, and where h_2 will bear the same relation to $f_2(\alpha)$. We shall then have two normal and orthogonal sets of functions $[h_1(T^n\alpha)]$ and $[h_2(T^n\alpha)]$, but there will not necessarily be any relation of orthogonality between these two sets.

Let us notice that if we put $F_{ij}(\theta)$ for the functions with Fourier coefficients

$$\int_0^1 f_i(T^n\alpha)f_j(\alpha)\,d\alpha, \tag{23}$$

[5] The condition can also be stated as necessary and sufficient as follows: The projection of $f(\alpha)$ on the closed linear manifold generated by the set of functions $\{f(T^{-n}\alpha), f(T^{-n-1}\alpha), \cdots\}$ converges in the mean to zero as n approaches infinity, if and only if the spectral distribution function is absolutely continuous and

$$\int_{-\pi}^{\pi} |\log F(\theta)|\,d\theta < \infty.$$

then

$$F_1(\theta) = F_{11}(\theta), \tag{24}$$

and

$$F_2(\theta) = F_{22}(\theta). \tag{25}$$

It is easy to prove that $F_{11}(\theta)$ and $F_{22}(\theta)$ are real and nonnegative while

$$F_{12}(\theta) = \bar{F}_{21}(\theta). \tag{26}$$

Moreover,

$$\begin{vmatrix} F_{11}(\theta) & F_{21}(\theta) \\ F_{12}(\theta) & F_{22}(\theta) \end{vmatrix} \tag{27}$$

can be shown to be nonnegative. Let us make the hypothesis

$$\int_{-\pi}^{\pi} \left| \log \begin{vmatrix} F_{11}(\theta) & F_{21}(\theta) \\ F_{12}(\theta) & F_{22}(\theta) \end{vmatrix} \right| d\theta < \infty. \tag{28}$$

Since we have made the supposition that the functions f_1 and f_2 belong to the class L_2, it is not difficult to prove that the functions $F_{ij}(\theta)$ all belong to the class L, so that the effective part of our assumption is

$$\int_{-\pi}^{\pi} \left| \log^- \begin{vmatrix} F_{11}(\theta) & F_{21}(\theta) \\ F_{12}(\theta) & F_{22}(\theta) \end{vmatrix} \right| d\theta < \infty. \tag{29}$$

Since however

$$\begin{vmatrix} F_{11}(\theta) & F_{21}(\theta) \\ F_{12}(\theta) & F_{22}(\theta) \end{vmatrix} = F_{11}F_{22} - F_{12}F_{21} = F_{11}F_{22} - |F_{12}|^2, \tag{30}$$

it will follow that

$$\int_{-\pi}^{\pi} |\log^- F_{11}(\theta)F_{22}(\theta)| \, d\theta < \infty \tag{31}$$

from which we may conclude that

$$\left. \begin{array}{l} \int_{-\pi}^{\pi} |\log F_{11}(\theta)| \, d\theta < \infty \\[2mm] \int_{-\pi}^{\pi} |\log F_{22}(\theta)| \, d\theta < \infty \end{array} \right\}, \tag{32}$$

which are the assumptions we have previously made separately for

$$F_{11}(\theta) \quad \text{and} \quad F_{22}(\theta).$$

3. [The successive projection technique]

I now wish to introduce a lemma of very general character concerning Hilbert space. It is the following:[6]

[Lemma I] *Let H_1 be a closed subspace of Hilbert space and let H_2 be another such closed subspace. Then their common part $H_1 H_2$ will be a closed subspace of Hilbert space. If f is any vector in Hilbert space, and if $P_1 f$ is the projection of f on H_1 while $P_2 f$ is the projection of f on H_2, then the result of consecutive projection*

$$P_1 f, \; P_2 P_1 f, \; P_1 P_2 P_1 f, \cdots,$$

will converge in the mean to the projection of f on $H_1 H_2$.

Let us note this H_1 contains two orthogonal spaces, one of which is $H_1 H_2$ while the other contains those functions in H_1 which are orthogonal to all functions in $H_1 H_2$. This other part we shall call H_1^*. Similarly, interchanging the roles of H_1 and H_2, we separate every function of H_2 into a part lying in H_1 and a part orthogonal to all functions in $H_1 H_2$ which we call H_2^*. Then the successive projection of a vector on H_1 and H_2 will be given by its projection on $H_1 H_2$ plus the result of its successive projection on H_1^* and H_2^*. H_1^* and H_2^* will not necessarily be orthogonal to one another, but they will at any rate contain no vector other than 0 belonging to both. If therefore I can prove that when I have two closed subspaces of Hilbert space H_1^* and H_2^* not containing any vector in common except 0, then the result of consecutive projection of these two will converge in the mean to 0, I shall have established my lemma.

Now let $\phi_n(\chi)$ be a set of normal and orthogonal functions belonging to H_1^* and closed on H_1^* and let $\psi_n(\chi)$ be a set of normal and orthogonal functions belonging to H_2^* and closed on H_2^*. Then if I start with any function $f(\chi)$ on H_1^*, I can write it

$$\sum A_n \phi_n(\chi). \tag{33}$$

---

[6] See von Neumann [1950, Theorem 13.7].

If I project this function on H_2^*, the projection will be

$$\sum_m \sum_n A_n \left[\int \phi_n \bar{\psi}_m \right] \psi_m(x); \tag{34}$$

projecting this back on H_1^*, I obtain

$$\sum_m \sum_n \sum_p A_n \left[\int \phi_n \bar{\psi}_m \int \psi_m \bar{\phi}_p \right] \phi_p(x). \tag{35}$$

The result of these repeated projections will be to change each function ϕ_n to

$$\sum Q_{pn} \phi_p, \tag{36}$$

where Q_{pn} will be

$$\sum_m \int \phi_n \bar{\psi}_m \int \psi_m \bar{\phi}_p. \tag{37}$$

That is, Q_{pn} will satisfy the condition that

$$Q_{np} = \bar{Q}_{pn}. \tag{38}$$

The operator of double projection will have Hermitian coefficients and will be what is known as a self-conjugate operator. It will also be an operator which reduces the length of any known nonzero vector in H_1^*.

Well-known theorems of Hermann Weyl prove that such an operator will have a spectrum continuous or discrete. To transform any function in H_1 by such an operator, we expand it in the spectral functions, and change each function by a factor which is less than one in absolute value. It is easy to prove that such an operator, when repeated sufficiently often, will turn any vector of finite length into a vector of length as small as we choose.

Let us apply this lemma to the two spaces H_1 and H_2 consisting respectively of all functions of L_2 orthogonal to the functions[7] $h_1(T^{-n}\alpha)$ and $h_2(T^{-n}\alpha)$. To form the projections of $h_1(\alpha)$ and $h_2(\alpha)$ on this space[8] is essentially the same thing as taking the projections of f_1 and f_2 respectively on spaces which are respectively dependent on f_1 and its past, but orthogonal to its past and dependent on f_2 and its past and orthogonal to that past. Let me start with h_1 and

[7] Where n is positive.
[8] That is, the common part $H_1 H_2$.

find an expression for the part of h_1 which is orthogonal to the past of f_1 and f_2 and form the part of h_2 which is orthogonal to the past of f_1 and f_2. These functions we shall call respectively $k_1(\alpha)$, $k_2(\alpha)$.

We shall have for the projection of h_1 orthogonal to its own past h_1 itself, and $h_1(\alpha)$ will be our first approximation in the mean to $k_1(\alpha)$. We shall now take the part of h_1 which will be orthogonal to the past of h_2. This will be

$$h_1(\alpha) - \sum_{m=1}^{\infty} h_2(T^{-m}\alpha) \int_0^1 h_1(\beta)\overline{h_2(T^{-m}\beta)} \, d\beta. \tag{39}$$

We project again to find the part orthogonal to the part of H_1 where h_1 is orthogonal to its past and will need no new term so that only the second term must be taken care of. It is clear that the extra added term to make the third approximation will be given by

$$+ \sum_{m=1}^{\infty} \sum_{n=1}^{\infty} h_1(T^{-n}\alpha) \int_0^1 h_1(\beta)\overline{h_2(T^{-m}\beta)} \, d\beta \int_0^1 h_2(T^{-m}\beta)\overline{h_1(T^{-n}\beta)} \, d\beta. \tag{40}$$

The rule of continuing this series is now clear, and the terms will alternately contain $h_1(\alpha)$, the past of $h_2(\alpha)$, the past of $h_1(\alpha)$, and so on. The signs of the terms will alternate. The coefficient of the first term will contain one integral and one sign of summation, that of the second two integrals and two signs of summation, and so on. This series

$$h_1(\alpha) - \sum_{m=1}^{\infty} h_2(T^{-m}\alpha) \int_0^1 h_1(\beta)\overline{h_2(T^{-m}\beta)} \, d\beta$$

$$+ \sum_{m=1}^{\infty} \sum_{n=1}^{\infty} h_1(T^{-m}\alpha) \int_0^1 h_2(T^{-n}\beta)\overline{h_1(T^{-m}\beta)} \, d\beta$$

$$\times \int_0^1 h_1(\beta)\overline{h_2(T^{-n}\beta)} \, d\beta$$

$$- \sum_{m=1}^{\infty} \sum_{n=1}^{\infty} \sum_{p=1}^{\infty} h_2(T^{-m}\alpha) \int_0^1 h_1(T^{-n}\beta)\overline{h_2(T^{-m}\beta)} \, d\beta$$

$$\times \int_0^1 h_2(T^{-p}\beta)\overline{h_1(T^{-n}\beta)} \, d\beta$$

$$\times \int_0^1 h_1(\beta)\overline{h_2(T^{-p}\beta)} \, d\beta$$

$$+ \cdots \qquad\qquad\qquad *$$

will be $k_1(\alpha)$. $k_1(\alpha)$ is then the part of $h_1(\alpha)$ which is orthogonal to the pasts of h_1 and h_2 so that

$$
\begin{aligned}
\int_0^1 |k_1(\alpha)|^2\, d\alpha &= \int_0^1 k_1(\alpha)\overline{k_1(\alpha)}\, d\alpha \\
&= \int_0^1 |h_1(\alpha)|^2\, d\alpha - \sum_{m=1}^{\infty} \left| \int_0^1 h_1(\alpha)h_2(\overline{T^{-m}(\alpha)})\, d\alpha \right|^2 \\
&\quad - \sum_{m=1}^{\infty} \left| \sum_{n=1}^{\infty} \int_0^1 h_1(\alpha)\overline{h_2(T^{-n}\alpha)}\, d\alpha \right. \\
&\quad \times \left. \int_0^1 h_2(T^{-n}\alpha)\overline{h_1(T^{-m}\alpha)}\, d\alpha \right|^2 \cdots.
\end{aligned}
\tag{41}
$$

Clearly

$$
\int_0^1 |k_1(\alpha)|^2\, d\alpha
\tag{42}
$$

is positive, and equally clearly

$$
\int_0^1 |h_1(\alpha)|^2\, d\alpha = 1.
\tag{43}
$$

Therefore

$$
\int_0^1 |k_1(\alpha)|^2\, d\alpha
\tag{44}
$$

lies between 0 and 1, and similarly

$$
0 \le \int_0^1 |k_2(\alpha)|^2\, d\alpha \le 1.
\tag{45}
$$

$k_1(\alpha)$ is that part of $h_1(\alpha)$ which is orthogonal to the pasts of both f_1 and f_2, while $k_2(\alpha)$ is that part of $h_2(\alpha)$ orthogonal to both pasts. Let us notice that

$$
\int_0^1 k_i(T^{-n}(\alpha))\overline{k_j(\alpha)}\, d\alpha
\tag{46}
$$

is always 0 if n is positive. From that and the measure-preserving character of T it results that

$$
\int_0^1 k_i(T^n\alpha)\overline{k_j(T^m\alpha)}\, d\alpha
\tag{47}
$$

is 0 unless m and n are the same. As yet, however, we know nothing in the case where m and n are the same, except that we may reduce this case to the case when both m and n may be given the value 0.

There are two cases which now present themselves. Either k_1 and k_2 have a relation of linear dependence[9] or they do not. If they are linearly independent, they cannot, either of them, be equivalent to 0. Let us suppose that k_1 is not equivalent to 0. Then we can normalize it to obtain $q_1(\alpha)$. We then form

$$k_2(\alpha) - q_1(\alpha) \int_0^1 k_2(\beta)\overline{q_1(\beta)}\, d\beta. \qquad (48)^*$$

This function is obviously orthogonal to q_1. If it is equivalent to 0, k_1, and k_2 are not linearly independent. If it is not equivalent to 0, it can be normalized, and thus we obtain $q_2(\alpha)$. Then the functions $q_1(\alpha)$ and $q_2(\alpha)$ are such that $q_i(T^m\alpha)$ form a normal and orthogonal set, any two of them being orthogonal, unless both i and m agree.

Continuing on the assumption that k_1 and k_2 are linearly independent, we can express f_1 and f_2 in terms of this normal and orthogonal set.[10] In proving this, we can establish that the formal series for $f_i(\alpha)$ is

$$\sum_{j=1,2} \sum_{n=0}^{\infty} q_j(T^{-n}\alpha) \int_0^1 f_i(\beta)\overline{q_j(T^{-n}\beta)}\, d\beta. \qquad (49)^*$$

By studying the partial sums of this series and the difference between these partial sums and $f_i(\alpha)$, we can see that either the series converges in the mean to $f_i(\alpha)$, or we shall have the projection of $f_i(\alpha)$

[9] The question of the independence of the functions k_1 and k_2 was originally expressed in quite different language by Zasuhin [1941] in terms of "the rank of the process." Wiener and Masani [1957, p. 145], using Zasuhin's terminology, identify the condition of "full rank" as being equivalent to the convergence of the logarithmic integral that appears in Theorem 2. The analytic characterization of the independence of k_1 and k_2 in terms of the logarithmic integral becomes apparent in the present proof just before the statement of Theorem 2.

[10] The basic hypotheses include the assumption that the projection of each f_1 and f_2 on the closed linear manifold generated by the two sets of functions, $\{f_1(T^{-n_1}\alpha), f_1(T^{-n_1-1}\alpha), \cdots\}$, $\{f_2(T^{-n_2}\alpha), f_2(T^{-n_2-1}\alpha), \cdots\}$, converges in the mean to zero as n_1 and n_2 approach infinity. After Kolmogorov [1941a], this condition has been called *regular* by Wiener and Masani [1957]. We may appeal to the theorem that was stated explicitly by Wiener and Masani [1957, p. 147]: A stationary process is regular and of full rank if and only if it has absolutely continuous spectral distribution and is of full rank. Using this terminology, the hypotheses of Theorem 2 below, as spelled out explicitly at the end of Section 2, are precisely the hypotheses of regularity and full rank.

on the remote past of f_1 and f_2 together not going to 0. Since the latter has been excluded, we shall have

$$f_i(\alpha) = \sum_{j=1,2} \sum_{n=0}^{\infty} q_j(T^{-n}\alpha) \int_0^1 f_i(\beta)\overline{q_j(T^{-n}\beta)}\, d\beta. \qquad (50)^*$$

Under these conditions

$$\int_0^1 f_i(T^m\alpha)\overline{f_j(\alpha)}\, d\alpha$$

$$= \sum_{K=1,2} \sum_{n=0}^{\infty} \int_0^1 f_i(T^{m+n}\beta)\overline{q_K(\beta)}\, d\beta \int_0^1 \overline{f_j(T^n\beta)}q_K(\beta)\, d\beta. \qquad (51)^*$$

Let us notice that the series

$$M_{ij}(\theta) = \sum_0^{\infty} e^{in\theta} \int_0^1 f_i(\beta)\overline{q_j(T^{-n}\beta)}\, d\beta \qquad (52)$$

will converge in the mean to functions belonging to L_2, and that [51] will [become]

$$\int_0^1 f_i(T^m\alpha)\overline{f_j(\alpha)}\, d\alpha$$

$$= \frac{1}{2\pi} \int_{-\pi}^{\pi} [M_{i1}(\theta)\overline{M_{j1}(\theta)} + M_{i2}(\theta)\overline{M_{j2}(\theta)}]\, e^{-im\theta}\, d\theta. \qquad (53)^*$$

In other words, if we use matrix notation, the matrices whose Fourier coefficients are given by the autocorrelation matrices with elements belonging to L can be factored into the matrix product

$$\mathbf{M}\cdot\tilde{\mathbf{M}}, \qquad (54)$$

where all the elements of $\mathbf{M}$ are the boundary values on the unit circle of functions of class L_2 analytic inside the unit circle, and indeed where it will not be difficult to show that the determinants of these matrices have no 0's inside the unit circle.

The other case which we have not yet discussed [11] is that in which

---

[11] It is at this point that further work needs to be accomplished. Though many authors, including Wiener and Masani [1957, 1958] and Wiener and Akutowicz [1959], have verified the spectral criteria for the matrix factorization by other means that do not include questions of convergence of series, a direct attack by the methods indicated here is still needed.

there is a linear relation between k_1 and k_2. If there is such a linear relation, at least one of the functions g_1 or g_2 can be expressed linearly in terms of the other and the past of both. In other words, we have a relation such as $f_1(\alpha) = cf_2(\alpha) +$ a vector in the past of f_1 and f_2.

Under these circumstances

$$\int_0^1 f_1(T^n\alpha)\overline{f_i(\alpha)}\, d\alpha = c \int_0^1 f_2(T^n\alpha)\overline{f_i(\alpha)}\, d\alpha \qquad (55)$$

plus something that may be approximated by a polynomial, always with the same coefficients, of the form

$$\int_0^1 f_2(T^{n-k}\alpha)\overline{f_i(\alpha)}\, d\alpha \qquad (56)*$$

and where the coefficients do not depend on n, but merely on i. It follows that if $\mathbf{H}(\theta)$ is the Hermitian matrix, of which the [Fourier transforms are autocorrelation coefficients], its elements will be such that

$$H_{1j}(\theta) = cH_{2j}(\theta) + \phi_1(\theta)H_{1j}(\theta) + \phi_2(\theta)H_{2j}(\theta), \qquad (57)$$

where ϕ_1 and ϕ_2 are free from singularity inside the unit circle. That is, the determinant

$$|\mathbf{H}(\theta)| \qquad (58)$$

will vanish indentically inside of the unit circle, and therefore, by a simple limit theorem, will vanish almost everywhere on the periphery. In other words, we have a situation which contradicts our assumption that

$$\int_{-\pi}^{\pi} |\log |\mathbf{H}(\theta)\|\, d\theta \qquad (59)$$

is finite. We may sum up these results in the following words.

[Theorem 2]   *If the Hermitian matrix* $\mathbf{H}(\theta)$ *has Fourier coefficients of the form*

$$\int_0^1 f_i(T^n\alpha)\overline{f_j(\alpha)}\, d\alpha, \qquad (60)$$

where f_1 *and* f_2 *belong to* L_2, *and if*

$$\int_{-\pi}^{\pi} |\log |\mathbf{H}(\theta)\|\, d\theta \qquad (61)$$

converges, then we may write

$$\mathbf{H}(\theta) = \mathbf{M}(\theta)\tilde{\bar{\mathbf{M}}}(\theta), \qquad (62)$$

where the elements of $\mathbf{M}$ *belong to* L_2 *inside any smaller circle concentric with the unit circle, and converge in the mean to their value on the unit circle. Indeed the determinant of the matrix* $\mathbf{M}$ *will be free from zeros inside the unit circle.*

4. [Extensions to Hermitian matrices of positive type]

We wish now to establish two further things: one, that any Hermitian matrix of positive type for which the integral of the logarithm of the determinant converges can be represented in the manner given above;[12] and second, that if the integral of the logarithm diverges, the matrix cannot be factored in the manner indicated. In order to establish the first of these results, let us suppose that a Hermitian matrix $\mathbf{H}$ can be written in the form

$$\mathbf{H}(\theta) = \mathbf{M}(\theta) \cdot \tilde{\bar{\mathbf{M}}}(\theta) \qquad (63)$$

where $\mathbf{M}$ is a matrix belonging to L_2. This is what we shall mean by saying that $\mathbf{H}$ is Hermitian and of positive type. I now introduce a variable α which I represent as before in binary form, but I now split its digits into two sequences labeled from $(-\infty, \infty)$ according to the rule

$$\alpha = .\alpha_1\, \alpha_2\, \alpha_3\, \alpha_4\, \alpha_5 \cdots$$
$$= .\beta_0\gamma_0\, \beta_1\gamma_1\, \beta_{-1}\gamma_{-1}\, \beta_2\gamma_2\, \beta_{-2}\gamma_{-2}\cdots. \qquad (64)$$

I write

$$B_n(\alpha) = 2\beta_n - 1; \qquad \Gamma_n(\alpha) = 2\gamma_n - 1. \qquad (65)$$

I introduce the transformation on α given by

$$T\alpha = .\beta_1\gamma_1\, \beta_2\gamma_2\, \beta_0\gamma_0\, \beta_3\gamma_3\, \beta_{-1}\gamma_{-1}\cdots. \qquad (66)$$

I put

$$M_{ij,n} = \frac{1}{2\pi}\int_{-\pi}^{\pi} M_{ij}(\theta)\, e^{-in\theta}\, d\theta. \qquad (67)$$

[12] The proof for this fact, outlined in the following, was circumvented by Wiener and Masani [1957], with the use of one of Cramér's theorems [1940]. The technique of the present proof has been studied in greater detail by Wiener and Akutowicz [1959].

I now define $f_i(\alpha)$ when i is one or two by

$$f_i(\alpha) = \sum_n (M_{i1,n}B_n(\alpha) + M_{i2,n}\Gamma_n(\alpha)). \tag{68}$$

Then it will not be difficult to prove that $\mathbf{H}(\theta)$ will have Fourier coefficients which can be written in the form

$$\int_0^1 f_i(T^n\alpha)\overline{f_j(\alpha)}\, d\alpha. \tag{69}$$

It remains to prove that if our logarithmic integral is infinite, no factorization can take place. However, if the factorization takes place and the said integral is infinite, then $\mathbf{M}(\theta)$ will exist such that all the elements will belong to L_2 and will be boundary values of functions analytic inside the unit circle and

$$\int |\log |\mathbf{M}(\theta)\|\, d\theta \tag{70}$$

are divergent. However, the determinant $|\mathbf{M}(\theta)|$ will be a function of L_2 around the unit circle and without zeros inside the unit circle, and we need only to appeal to our scalar theorem to show the impossibility of the vector situation.

5. Multiple prediction

As is well known, the linear prediction of single series stationary processes originated with the simultaneous work of Wiener [1949a] and Kolmogorov [1941a, 1941b] (the former being available for some years only as a classified document). Wiener's original work on prediction relied heavily upon the analysis of L_2-functions that vanish on the half-line (see Paley and Wiener [1934]). The work of Kolmogorov was directly related to the earlier analysis of Wold [1938] on stationary time series.

The probability space that underlies the work of this section will be $(A, \mathfrak{B}, \mu)$, where $A \equiv [0, 1]$ is the unit interval, $\mathfrak{B}$ is the corresponding Borel field, and where the probability measure μ corresponds to the ordinary Lebesgue measure.

Let $f_i(\alpha)$, $i = 1, 2, \cdots, N$, be complex measurable functions belonging to class L_2, and let T be a one-to-one point transformation (measurable and measure-preserving) carrying the unit interval onto itself. The time series under consideration will be the complex stochastic processes $\{f_i(T^n\alpha); n = \cdots -1, 0, 1, \cdots\}$ with the index n, the power of the transformation, corresponding to increment steps on the time axis. For convenience we will take $n = 0$ as the present and associate negative powers of T with the past.

The "multiple prediction" of the complex random variable, $f_i(T^m\alpha)$, $m > 0$, will be some function of the present and past values of all of the $f_i(\alpha)$, $i = 1, 2, \cdots, N$, which is close in some sense to the value of $f_i(T^m\alpha)$. The criterion of "closeness" is conventionally defined in terms of mean-square deviation from the true value, and for expedience's sake only the closest among all *linear* combinations of the present and past are considered. Thus, *we will assume that the linear multiple prediction of* $f_i(T^m\alpha)$ *will be some linear combination of the form*

$$L_i^{(m)}(\alpha) = \sum_{j=1}^{N} \sum_{n=0}^{\infty} f_j(T^{-n}\alpha)P_n^{(m)}(i, j), \qquad (71)$$

with convergence defined in the L_2-metric, and the "best" among all these linear forms will be that which minimizes the following:

$$\int_0^1 |f_i(T^m\alpha) - L_i^{(m)}(\alpha)|^2 \, d\alpha. \qquad (72)$$

The minimum value of the integral 72, $[\sigma_i^{(m)}]^2$, is the prediction error.

A clarification of the conditions on the spectrum under which the predictor can be represented in the form appearing in Equation 71 in the single series case was indicated by Wiener [1949b] and was completed by Akutowicz [1957]. The work of Wiener and Masani [1958] has further clarified the needed spectral conditions in the multiple series case.

The following assumptions (together with the assumptions on the meaningfulness of the forms 71 and 75) are sufficient for the present solution in which the $P_n^{(m)}(i, j)$ that minimize the integral 72 are obtained from a sample value of the full pasts. The assumptions i and ii to follow could be replaced by the requirements of regularity and full rank that were used in the matrix factorization of the last four sections (see footnotes 9 and 10).

i. *The projection of each $f_i(\alpha)$, $i = 1, 2, \cdots, N$, on the closed linear manifold generated by the set of functions*

$$f_1(T^{-n_1}\alpha), \ f_1(T^{-n_1-1}\alpha), \cdots,$$
$$f_2(T^{-n_2}\alpha), \ f_2(T^{-n_2-1}\alpha), \cdots,$$
$$\vdots \qquad\qquad \vdots$$
$$f_N(T^{-n_N}\alpha), \ f_N(T^{-n_N-1}\alpha), \cdots,$$

converges in the mean to zero as $n_1, n_2, \cdots, n_N$ approach infinity.

ii. *Let c_i, $i = 1, 2, \cdots, N$, be any N complex numbers not all zero. Then*

$$\int_0^1 \left| \sum_{j=1}^N c_j f_j(T\alpha) - L(\alpha) \right|^2 d\alpha > 0$$

for any linear combination $L(\alpha)$ of the form 71.

It is well known (see Chapter 1) that in the single series case assumption i implies assumption ii.

Property ii is familiar in the single series case. It states that the prediction error is positive and is often expressed analytically as the convergence of a logarithmic integral. The present form is more convenient for our purposes and explicitly excludes the possibility of any $f_i(\alpha)$ being a linear combination of its own past and the present and past of the other series. Assumption i, when expressed in the usual form for one series, states that the spectral distribution function is absolutely continuous and the logarithmic integral referred to converges. The wording in terms of the spectral distribution, which does not appear explicitly in our solution, is avoided here, though full details are available in a number of sources (Wiener and Masani [1957, 1958]). The purpose of assumption i is to exclude from the mathematics those time series containing a deterministic component.

It is obvious that the time series under consideration, by virtue of their construction through the transformation T, are strictly stationary. This is a specialization that can be avoided. For the present, the intuitive advantages of the specialization are important to us, as are the advantages of emphasizing the close relationship between strictly stationary time series and certain entities in statistical mechanics. For a careful discussion of the generality with which a stationary time series can be expressed in the present form, the reader is referred to Doob [1953, pp. 452–457].

Once the formulation of time series in terms of a measurable and measure-preserving transformation has been made, it is immediately possible to make a direct comparison between the series and dynamical flows in Gibbsian phase space. A point α can be identified with the state of a physical system of energy E expressed in terms of the Hamiltonian variables, T can be identified with the dynamics (i.e., the Hamiltonian equations) that transforms phase space onto itself with the passage of time, and, in this context, f_i can be identified with some phase functions of the system. The measure-preserving property of T is suggested by the Liouville theorem of statistical mechanics and, while it is not apparent how the physically motivated concept of metric transitivity (or ergodicity) enters into prediction theory, the following may be observed. The solutions that result from the linear prediction theory are in terms of the "phase average" form of the covariances

$$\int_0^1 f_i(T^{-\nu}\alpha)\overline{f_j(\alpha)}\, d\alpha.$$

If the ergodic assumption is invoked for the transformation T, then from the corollary of the Birkhoff theorem it is apparent that a sample value of the full past of each series will suffice to determine the covariances in the "time average" form

$$\lim_{N\to\infty} \frac{1}{N} \sum_{k=1}^{N} f_i(T^{-\nu-k}\alpha)\overline{f_j(T^{-k}\alpha)}.$$

In fact, it is only rarely that one can measure the covariances directly by averaging in "phase." Thus, it is evident that two of the same basic properties that are so well known in statistical mechanics appear now in the application of prediction theory.

The objective of our discussion is to express each $f_i(T^m\alpha)$, $i = 1$, $2, \cdots, N$, $m > 0$, as the sum of two parts

$$f_i(T^m\alpha) = F_i^{(m)}(\alpha) + G_i^{(m)}(\alpha), \tag{73}$$

where $F_i^{(m)}(\alpha)$ is linearly dependent on the present and past of all series or, in other words, expressible in the form 71, and where

$G_i^{(m)}(\alpha)$ is orthogonal to the present and past of all series or, in symbols:

$$\int_0^1 G_i^{(m)}(\alpha)\overline{f_j(T^{-n}\alpha)}\,d\alpha = 0, \qquad j = 1, 2, \cdots, N, \qquad n = 0, 1, 2, \cdots.$$

If we succeed in this, $F_i^{(m)}(\alpha)$ will be the unique closest point to $f_i(T^m\alpha)$ in the manifold generated by the present and past of all series, since $f_i(T^m\alpha) - F_i^{(m)}(\alpha) = G_i^{(m)}(\alpha)$ is orthogonal to that manifold. The measure of closeness is, of course, taken in the sense of the mean square. In other words, $F_i^{(m)}(\alpha)$ will be the best linear multiple prediction of $f_i(T^m\alpha)$ as already defined.

Throughout the remainder of this section we will use the conventional inner product notation. See Section 6 of the Introduction. All the infinite sums and the limits will be defined in terms of convergence in the L_2-metric, unless otherwise indicated.

The basic lemma that stands at the foundation of the present approach to multiple prediction is the same that characterized the technique of the matrix factorization. The lemma was stated and proved in Section 3.

If we define H_i, $i = 1, 2, \cdots, N$, such that its orthogonal complement $H_i^\perp$ relative to L_2-space is that closed linear manifold generated by the past of $f_i(\alpha)$, i.e.,

$$\{f_i(T^{-n}\alpha), \qquad n = 1, 2, \cdots\},$$

then by the lemma, the result of consecutive projections

$$P_1 f_1(\alpha), \ P_2 P_1 f_1(\alpha), \ P_1 P_2 P_1 f_1(\alpha), \cdots,$$

will converge in the mean to the function $g_{1,2}(\alpha)$, that part of $f_1(\alpha)$ that is orthogonal to its own past and to the past of $f_2(\alpha)$. Symbolically, the preceding statement can be written as follows:

$$g_{1,2}(\alpha) = \lim_{M \to \infty} [P_2 P_1]^M f_1(\alpha).$$

Similarly,

$$g_{2,2}(\alpha) = \lim_{M \to \infty} [P_2 P_1]^M f_2(\alpha)$$

will be that part of $f_2(\alpha)$ orthogonal to its own past and to the past of $f_1(\alpha)$.

Now if the intersection $H_1 H_2$ is chosen as one subspace and H_3

as another, then from the lemma it also follows that the result of successive projections

$$\cdots P_1 P_2 P_1 f_1(\alpha), \ P_3 \cdots P_1 P_2 P_1 f_1(\alpha), \cdots P_1 P_2 P_1 P_3 \cdots P_1 P_2 P_1 f_1(\alpha), \cdots$$

will converge in the mean to the projection of $f_1(\alpha)$ on the intersection $H_1 H_2 H_3$. In other words, we will obtain a function $g_{1,3}(\alpha)$ which is that part of $f_1(\alpha)$ orthogonal to its own past and to the pasts of $f_2(\alpha)$ and $f_3(\alpha)$. The limit of these consecutive projections can be written

$$g_{1,3}(\alpha) = \lim_{M_2 \to \infty} \left[P_3 \lim_{M_1 \to \infty} [P_2 P_1]^{M_1} \right]^{M_2} f_1(\alpha).$$

In general, the part of $f_i(\alpha)$ that is orthogonal to the pasts of $f_j(\alpha), j = 1, 2, \cdots, N$, can be written

$$g_{i,N}(\alpha) = \lim_{M_N \to \infty} \left[P_N \cdots \lim_{M_2 \to \infty} \left[P_3 \lim_{M_1 \to \infty} [P_2 P_1]^{M_1} \right]^{M_2} \cdots \right]^{M_N} f_i(\alpha). \quad (74)$$

For simplicity, from now on we will suppress the subscript N in $g_{i,N}(\alpha)$. *Let us assume that $g_i(\alpha)$ can be put in the form*[13]

$$g_i(\alpha) = f_i(\alpha) d_0(i, i) + \sum_{j=1}^{N} \sum_{n=1}^{\infty} f_j(T^{-n}\alpha) d_n(i, j), \quad (75)$$

where $d_n(i, j)$ are complex numbers, and

$$(f_i, g_j T^n) = 0, \qquad n > 0, \qquad i, j = 1, 2, \cdots, N. \quad (76)$$

From assumption ii and the form 75 it follows immediately that

$$\left\| \sum_{j=1}^{N} e_j g_j \right\|^2 > 0$$

for any combination of complex numbers $e_j, j = 1, 2, \cdots, N$, not all zero. Thus, the set $\{g_j, j = 1, 2, \cdots, N\}$ can be orthonormalized. With the Gram-Schmidt orthonormalization procedure, there is a set of coefficients, $b_{i,j}$, such that

$$q_i(\alpha) = \sum_{j=1}^{i} b_{i,j} g_j(\alpha) = \sum_{j=1}^{i} f_j(\alpha) a_0(i, j) + \sum_{j=1}^{N} \sum_{n=1}^{\infty} f_j(T^{-n}\alpha) a_n(i, j),$$

$$(77)$$

$$(q_i, q_j) = \begin{cases} 1, & i = j, \\ 0, & i \neq j. \end{cases}$$

[13] Spectral conditions that ensure the meaningfulness of this assumption originated with Wiener [1949b]. Specifically, see Wiener and Masani [1958].

It follows that

$$(q_i, q_j T^{-n}) = \begin{cases} 1, & i = j, \quad n = 0, \\ 0, & \text{otherwise,} \end{cases}$$
$$(f_i, q_j T^n) = 0, \qquad n > 0,$$
$$(f_i, q_j) = 0, \qquad i < j.$$

From assumption i, each $f_i(\alpha)$, $i = 1, 2, \cdots, N$, can be expanded in terms of the orthogonal set $\{q_j(T^{-n}\alpha); \; j = 1, 2, \cdots, N, \; n = 0, 1, 2, \cdots\}$,

$$f_i(\alpha) = \sum_{j=1}^{N} \sum_{n=0}^{\infty} q_j(T^{-n}\alpha)(f_i, q_j T^{-n}). \tag{78}$$

That is to say, $f_i(T^m\alpha)$, $i = 1, 2, \cdots, N$, $m > 0$, can be expressed in the desired form 73, where

$$F_i^{(m)}(\alpha) = \sum_{j=1}^{N} \sum_{n=m}^{\infty} q_j(T^{m-n}\alpha)(f_i, q_j T^{-n}), \tag{79}$$

$$G_i^{(m)}(\alpha) = \sum_{j=1}^{N} \sum_{n=0}^{m-1} q_j(T^{m-n}\alpha)(f_i, q_j T^{-n}). \tag{80}$$

With this the prediction problem is theoretically solved, for expressions 77 and 79 give the prediction in the form 71:

$$F_i^{(m)}(\alpha) = \sum_{j=1}^{N} \sum_{n=0}^{\infty} f_j(T^{-n}\alpha) \sum_{v=0}^{n} \left[\sum_{J=1}^{N} a_{n-v}(J, j)(f_i, q_j T^{-v-m}) \right]. \tag{81}$$

The expressions 72 and 80 give the prediction error

$$[\sigma_i^{(m)}]^2 = \sum_{j=1}^{N} \sum_{n=0}^{m-1} |(f_i, q_j T^{-n})|^2. \tag{82}$$

The coefficients a_n in the right-hand side of Equation 81 are obtainable from a set of *scalar* recurrence relations that follow from Equation 77 and the properties of the $q_i(\alpha)$

$$\left. \begin{array}{l} a_0(i, j) = 0, \qquad i < j, \\[2mm] \displaystyle\sum_{j=k}^{i} (f_j, q_k)a_0(i, j) = \begin{cases} 1, & k = i, \\ 0, & k \neq i, \end{cases} \\[4mm] \displaystyle\sum_{j=1}^{N} \sum_{n=0}^{v-1} (f_j, q_k T^{n-v})a_n(i, j) + \sum_{j=k}^{N} (f_j, q_k)a_v(i, j) = 0, \qquad v > 0. \end{array} \right\} \tag{83}$$

These relations give $a_n(i,j)$ in terms of $\{(f_k, q_l T^{-m}); \ m \geq 0,$ $k, l = 1, 2, \cdots, N\}$. It remains to express $(f_i, q_j T^{-n})$ in terms of the covariances $\{(f_k, f_l T^{-m}); \ m \geq 0, \ k, l = 1, 2, \cdots, N\}$.

Returning to the construction of the $q_i(\alpha)$, let us define as in Section 1 (see footnote 3),

$$h_i(\alpha) = \frac{P_i f_i(\alpha)}{\|P_i f_i\|}, \qquad i = 1, 2, \cdots, N,$$

to be the innovations for the respective series. These are linearly dependent on the present and past of $f_i(\alpha)$, $i = 1, 2, \cdots, N$, respectively, and represent the part of $f_i(\alpha)$ that is orthogonal to its own past. Therefore, with assumption i each set

$$\{h_i(T^{-n}\alpha), \qquad n = 0, 1, 2, \cdots\}, \qquad i = 1, 2, \cdots, N,$$

represents a complete and orthogonal set for the corresponding $f_i(\alpha)$.

$$\left.\begin{array}{l} (h_i T^{-n}, h_i T^{-m}) = \begin{cases} 1, & n = m, \\ 0, & n \neq m, \end{cases} \\[2ex] f_i(\alpha) = \sum_{n=0}^{\infty} h_i(T^{-n}\alpha)(f_i, h_i T^{-n}). \end{array}\right\} \qquad (84)$$

If we write

$$k_i(\alpha) = \frac{g_i(\alpha)}{\|P_i f_i\|}, \qquad (85)$$

then, in terms of the innovations, the $k_i(\alpha)$ take the form

$$k_i(\alpha) = \lim_{M_N \to \infty} [P_N \cdots \lim_{M_2 \to \infty} [P_3 \lim_{M_1 \to \infty} [P_2 P_1]^{M_1}]^{M_2} \cdots]^{M_N} h_i(\alpha). \quad (86)$$

If we write $P_i^\perp$ for the projection on $H_i^\perp$ so that $P_i = 1 - P_i^\perp$, then P_j operating on $h_i(\alpha)$ can be expressed analytically as $h_i(\alpha)$ minus the development of $h_i(\alpha)$ in terms of the basis $\{h_j(T^{-n}\alpha); n = 1, 2, \cdots\}$ of the closed linear manifold generated by the past of $f_j(\alpha)$.

The functions $q_i(\alpha)$, as defined by Equation 77, can be interpreted as the orthonormalization of the $k_i(\alpha)$:

$$q_i(\alpha) = \sum_{j=1}^{i} b'_{i,j} k_j(\alpha), \qquad (87)$$

where $b'_{i,j}$ are determined by the constants

$$(k_i, k_j) = (h_i, k_j), \qquad i, j = 1, 2, \cdots, N. \qquad (88)$$

From the representations 84 and 87 we see that the inner products which we wish to compute have the form

$$(f_i, q_j T^{-n}) = \sum_{l=1}^{j} b'_{j,l}(f_i, k_l T^{-n})$$

$$= \sum_{l=1}^{j} b'_{j,l} \sum_{m=0}^{\infty} (f_i, h_i T^{-m})(h_i T^{-m}, k_l T^{-n}). \qquad (89)$$

From Equations 86, 88, and 89 it is clear that the desired inner products $(f_i, q_j T^{-n})$ involve only $\{(f_i, h_i T^{-m}); \ m \geq 0\}$ and $\{(h_k, h_l T^{-m}); \ m \geq 0, \ k, \ l = 1, 2, \cdots, N\}$. Hence it remains to express the latter inner products in terms of the covariances.

It can be shown in the one series case that if we define in $L_2 -$ norm

$$\psi_i(\theta) = \sum_{n=0}^{\infty} e^{in\theta}(f_i, h_i T^{-n}), \qquad i = 1, 2, \cdots, N, \qquad (90)$$

then the analytic function

$$\psi_i(r, \theta) = \sum_{n=0}^{\infty} r^n e^{in\theta}(f_i, h_i T^{-n}), \qquad 0 \leq r < 1,$$

has the property that its phase can be determined from a knowledge of the absolute value of $\psi_i(\theta)$, that is for $0 \leq r < 1$,

$$\log \psi_i(r, \theta)$$
$$= \frac{1}{2\pi} \left\{ \sum_{n=1}^{\infty} r^n e^{in\theta} \int_{-\pi}^{\pi} \log |\psi_i(x)|^2 \, e^{-inx} \, dx + \int_{-\pi}^{\pi} \log |\psi_i(x)| \, dx \right\}, \qquad (91)$$

and moreover,

$$\frac{1}{2\pi} \int_{-\pi}^{\pi} |\psi_i(\theta)|^2 \, e^{in\theta} \, d\theta = (f_i T^{-n}, f_i). \qquad (92)$$

Thus, if we also define, as in Section 3,

$$\phi_{i,j}(\theta) = \sum_{n=-\infty}^{\infty} e^{in\theta}(f_i, f_j T^{-n})$$

$$= \sum_{n=0}^{\infty} e^{in\theta}(f_i, f_j T^{-n}) + \sum_{n=1}^{\infty} e^{-in\theta}(f_i T^{-n}, f_j), \qquad (93)$$

$i, j = 1, 2, \cdots, N$, for which it can be found that

$$\frac{\phi_{i,j}(\theta)}{\psi_i(\theta)\overline{\psi_j(\theta)}} = \sum_{n=-\infty}^{\infty} e^{in\theta}(h_i, h_j T^{-n}), \tag{94}$$

it is clear how a knowledge of the covariances $\{(f_i, f_j T^{-n}); n \geq 0,$
$i, j = 1, 2, \cdots, N\}$ is used to determine the prediction.

As mentioned earlier, the metric transitivity assumption makes a
knowledge of sample values of the full pasts sufficient to predict the
future and a knowledge of the "phase average" form of the co-
variance is unnecessary.

5

The Differential-Space Theory of Quantum Systems

I. The question of "hidden variables" in quantum mechanics

Since the very earliest period, the physical interpretation of quantum mechanics has been a subject of controversy. As the discussion progressed, it gradually became centered around the conflict between two well-defined opposite points of view.[1] Although hitherto neither has obtained a clear-cut logical victory, most physicists have tended to resolve their doubts in favor of the one identified mainly with Niels Bohr. Bohr adapts the physical picture to the schematic experimental situation, excluding any concepts that do not correspond strictly to experimental elements, even if there is no actual disagreement involved. Thus, in a situation in which the position of a particle is sharply defined, the very concept of momentum is excluded (as a physical reality, although retained in a formal sense) since the wave function displays a wide dispersion of momentum values.

The principal advocate of the opposing view has been Albert Einstein. To use Einstein's characterization, quantum mechanics (which may be considered a formal expression of Bohr's point of view) is not a "complete theory"; elements of physical reality, such as momentum in the above example, may exist and have definite

---

[1] The history of this controversy is well summarized, with references, in the collection of essays edited by Schilpp [1949]; see especially those by Bohr (pp. 201–241) and Einstein (pp. 665–688). Other particularly important references are Einstein, Podolsky, and Rosen [1935], Bohr [1935], Furry [1936], Einstein [1936], and Feyerabend [1962].

values, even though not sharply defined according to the wave function.

Since advocates of the first view maintain that the quantum-mechanical description of reality is complete, it would follow that each individual of a number of identically prepared maximally defined systems (that is, subject to the indeterminacy relations) is *completely* represented as to its physical state by the same wave function. The fact that a measurement of a variable of which this wave function is not an eigenfunction will yield a variety of values for different individual systems is not, according to Bohr, to be used as an argument for supposing that there exists a theoretically definable difference between these systems prior to the measurement. Indeed, such a conclusion is easily shown not to be necessary. The variable that was sharp in the wave function and the variable finally measured correspond to mutually exclusive possibilities of definition experimentally; the process of measuring the latter includes an unobservable and uncontrollable disturbance of the value of the former, making its value different[2] in each case. This accounts for the difference between systems after the measurement, without assuming a difference before measurement. But Bohr replaces this negative conclusion (lack of necessity for assuming a theoretically definable difference prior to measurement) by a positive one. The differences between different systems with regard to the second variable are, so to speak, created at the moment of its measurement and did not exist previously. Not that all had identical values; rather, one must say that the second variable (within more or less broad limits depending on the particular wave function) has no clear-cut prior existence; not only its values, but the second variable as a defined property, are first created by the act of measuring it. The logic of the assertion that the systems are different after the measurement, even though identical before, is protected by the fact of the unobservable disturbance of the first variable; this is the essential role of the indeterminacy principle.

Without questioning the ability of these concepts to account for the phenomena, Einstein insisted that it had not been proved that

---

[2] Actually this variable becomes *undefined* after the measurement, as was the other before; see what follows.

physical theory is constrained from examining the area of indeterminacy which Bohr maintains to be inscrutable even in concept. He asserted, in effect, that when there exists a difference between measured values of a variable in two systems, there exists a difference between them which is physically definable, in principle. It is not necessary to deny the undeniable disturbance sometimes accompanying the act of measurement in order to assert this; indeed, the difference can be demonstrated within the very context of quantum mechanics itself, even when there is no disturbance (Einstein, Podolsky, and Rosen [1935]). Einstein in effect challenged Bohr's transformation, described above, of a negative into a positive statement.

As we shall develop it in the following discussion, the differential-space theory of quantum systems (Wiener and Siegel [1953, 1955]; Siegel and Wiener [1956]) carries the controversy sketched previously to a new stage of concreteness. This is because it realizes the formal program of Einstein; namely, a physical picture in which a collection of identically prepared systems (that is, subject to the indeterminacy principle) is represented by an ensemble of different, precisely defined systems so that the dispersion in values of measured quantities has its counterpart in the variety of properties of the systems in the ensemble. On the other hand, the theory and current experimental information are in too provisional a state for these facts to constitute in any sense a refutation of Bohr's point of view. That is, it is entirely possible that the individual-description concept formally furnished by the differential-space theory may turn out to be physically vacuous.

2. Individual and ensemble

It is a time-honored fact that assigning (in imagination) sharp values to two conjugate variables of a system, and doing nothing more, will result in conflict with the experimental facts. In a two-slit diffraction experiment with electrons (see, for example, L. I. Schiff [1955]) the positional undefinability-in-principle of the electron in the Bohr picture serves an important logical purpose. If one could think of the electron as defined within a region of extent less than the separation of the slits (assuming it to be experimentally not so well defined), one would be able to assert (in thought) that the electron had certainly gone through one of the slits, and one would therefore be unable to account for the influence of the other slit on the experimental results. (The results show a diffraction pattern of structure dependent on the separation of the slits, since the electrons are, by hypothesis, experimentally so prepared as to give such a pattern. In particular, their experimental range of position definition is broader than the separation of the slits.)

Referring to the first sentence of the preceding paragraph, it is then evident that in order to realize a description of quantum systems in which all observables are defined one must do "something more." The classic way to do this "something more" is to add further determining—"hidden"—variables to the system; these can then serve the all-important purpose of expressing the influence of "this slit" even when the particle is passing through "the other slit."

A theory of hidden variables designed to explain quantum mechanics must contain two basic features. One is the definition of the hidden variables themselves, which furnish a complete description of a unique system. This may be called the *dynamical* part of the theory. The other part of such a theory is its *statistical* part. This has the following function: to show how ensembles of completely described systems obeying the postulated dynamics can be so constructed as to have the same statistical properties as are expressed by a given quantum-mechanical wave function. If nature is correctly described by such a theory, individual events as described by the conventional quantum mechanics are *crypto-deterministic* (the term is due to Whittaker), that is, deterministic in themselves, but with an imprecision in their definition corresponding to an imprecision in their experimental measurement.

The program stated in the preceding paragraph implies certain necessary conditions on its realization, tied up with the intuitive meanings our minds attach to the terms "deterministic" and "statistical." The following is what we feel to be a "fair" or "reasonable" formulation of these conditions.

1. The individual system is deterministic, i.e., its present state (the present values of all measurable variables, including the hidden ones, which must be assumed measurable in principle, although not yet in fact) determines its future state (the values of all measurable variables at any future time). The system must of course be closed if this is to apply. Open systems must first be combined with all systems with which they interact (or be subjected to some equivalent procedure), making them closed, before they can be deterministic.

The foregoing condition concerns only what we have called the *dynamical* part of the theory. The remaining two conditions refer to the *statistical* part. It is assumed, following Gibbs, that all statistical predictions, that is, predictions regarding the distribution of values of physical properties over a large number of identically, but not (in the dynamical sense) precisely, defined systems, are to be obtained from an ensemble of dynamically precisely defined systems, such that a measure (or probability, or weight) is given for any suitably defined and restricted set of values of the variables. This measure is assumed to be the probability, relative to the ensemble, of the subensemble of all systems belonging to the ensemble and having values of their variables in this set. Since the ensemble is a mathematical representation of the set of actual systems observed, and one cannot isolate a negative number of systems, one infers the second condition:

2. The measure of a suitably defined subensemble is never negative.[3]

The correspondence between the ensemble and the set of actual observed systems is presumed to be preserved with the passage of time. That is, the systems of the ensemble duplicate the time-behavior of the actual observed systems. The third condition is the formal expression of this idea.

3. The measure of a subensemble corresponding to a given set of variables is conserved as this set changes according to the natural motion (dynamics) of the variables contained in it. This expresses

---

[3] Ensembles agreeing with quantum mechanics, but having negative measure for some subsets, have been defined. See E. P. Wigner [1932]; J. P. Moyal [1949]; H. Weyl [1931], pp. 274, 275. Such an ensemble cannot have a collection of real physical systems as its counterpart.

the conservation of probability, or preservation of measure, in time; in classical statistical mechanics, the proof of this theorem rests on Liouville's theorem.

As our analysis of the differential-space theory of quantum systems develops in this chapter, we shall show how the theory (in our opinion) satisfies the spirit of these conditions, and to what extent it is necessary to modify their letter in fitting them to the peculiar conditions of quantum phenomena. It should be emphasized that this theory was contrived for the rather special purpose of reducing the apparently anomalous probabilistic assertions of quantum mechanics[4] to ordinary probability methods, this being accomplished by the introduction of the hidden variables. No attempt has been made to introduce any new physical content apart from the conceptual replacement of probability amplitudes by probabilities. Thus any ultimate physically interesting formulation of hidden variables may well differ considerably from this one. If this is true, it may be that the latter is to be regarded mainly as a means of proving, by construction, the possibility of a theory of hidden variables consistent with quantum phenomena.[5]

3. Résumé of quantum mechanics

The differential-space quantum theory makes use of the entire *mathematical* apparatus of quantum mechanics. Thus it must not be thought of as replacing this apparatus, but rather as enlarging it somewhat (through the addition of the new mathematical element of differential space). On the other hand, it does reject the traditional statement of the *statistical* content of quantum mechanics. By "statistical content" we mean the postulates whereby the frequencies of events to be expected on performing certain types of experiments are inferred. These postulates are replaced by postulates adapted to the nature of differential space.

[4] By "quantum mechanics" we shall always mean the standard theory, as fundamentally expounded in, say, the works of Dirac [1958] (for physicists) or von Neumann [1932b] (better adapted to the mathematician's taste).

[5] The work of Bohm [1952, 1957] may be regarded also as such an existence theorem, and is prior to the differential-space theory. The two theories differ considerably in their nature.

From the foregoing, it follows that we shall have to summarize the mathematics of quantum mechanics, in order to obtain a basis for the differential-space theory. Moreover, we shall have to state the statistical postulate of quantum mechanics in order (*a*) to see what is being replaced and (*b*) to ensure the physical correctness of the differential-space theory, which can at the present stage be most readily done by requiring the frequencies of events to agree with the experimentally proved frequencies derived from quantum mechanics. (This does not close the door to eventual disagreement between the two theories when the differential-space theory is further developed, but the disagreement must not concern experiments of now-known type, apart possibly from minute effects not now measurable.)

We shall regard a physical system as an *individual* belonging to a *species*. We confine ourselves to so-called "ordinary" quantum mechanics, i.e., to systems composed of point-mass particles or at any rate to systems having a finite number of degrees of freedom. Then a species is determined by the numbers, masses, and intrinsic spins of the particles present, plus their mutual forces and the external forces to which they are subjected. Its individuals will differ with respect to the values found on measurement of coordinates, momenta, energy, spin quantum numbers, etc., of the individual particles.

The mathematical basis of quantum mechanics is now easily stated: It is the Hilbert space of L_2 functions of the space and spin coordinates (case of systems of mass particles), or of any equivalent set of generalized coordinates, appropriate to the description of the species of system considered. The maximum amount of experimental information obtainable about a given physical system is assumed to determine a unique element of infinite-dimensional Hilbert space $\mathfrak{H}_\infty$. This element contains, in a way we shall describe, all information concerning statistical frequencies of all possible results of each and every kind of physical observation performable on the system. In the original wave theory of Schrödinger this element was a function of three-dimensional physical space, called a "wave function." This term is often retained even when other dimensions and elements are added to the set of the variables of the function.

Each physical observable of a system is correlated with a hyper-maximal Hermitian operator[6] in $\mathfrak{H}_\infty$ (von Neumann [1932b, p. 88]). A hypermaximal operator is one whose eigenvalue problem is soluble. Let us specialize for a while to a Hilbert space of functions of a single variable. (The general case will be taken up at the end of Section 5.) A hypermaximal operator, say $\mathbf{R}$, gives rise to a unique "resolution of unity," that is, a family of projection operators[7] $\mathbf{E}(r)$, $-\infty < r < \infty$, such that $\mathbf{R}$ may be written in the symbolic form

$$\mathbf{R} = \int_{-\infty}^{\infty} r \, d\mathbf{E}(r), \tag{1}$$

in the sense that for all elements f, g of $\mathfrak{H}_\infty$ one has

$$(f, \mathbf{R}g) = \int_{-\infty}^{\infty} r \, d(f, \mathbf{E}(r)g), \tag{2}$$

the right-hand side being a Stieltjes integral. For further details of the definition of $\mathbf{E}(r)$, see von Neumann [1932b, p. 61].

In order to make contact with Dirac's presentation, which is one of the classic presentations of quantum mechanics, and will be useful in furnishing *façons de parler* for our discussion, we shall digress here to give a brief summary of his characteristic notation. He emphasizes the vector character of an element f of $\mathfrak{H}_\infty$ (associated with a function $f(g)$, say) by writing it $|f\rangle$. With $|f\rangle$ is associated its dual $\langle f|$ (called "complex conjugate"—this is a generalization of the

---

[6] Due to the characteristic statistical postulates of quantum mechanics, the observable and operator are in a one-to-one relationship, and the distinction between them can often be overlooked, at least verbally; hence in conventional treatments of quantum mechanics (see Dirac [1958]) it is customary to apply the word "observable" to the operator itself. But in the differential-space theory, while the association between operator and observable is retained, the way of assigning values of the observable to individual systems is radically different from that of quantum mechanics. Thus we shall always retain the distinction in meaning between operator and observable.

[7] For the sake of convenience and suggestiveness of notation, we use bold-face capitals for operators, and ordinary capitals for observables (which are also to be considered random variables). For the realized value or as a variable of integration corresponding to an observable R, or operator $\mathbf{R}$, we use the lower case of the same letter, viz. r.

usual meaning). Vectors $|f\rangle$ are called *kets*, $\langle f|$ are called *bras*. The inner product (f, g) is written

$$\langle f \,|\, g \rangle. \tag{3}$$

The latter is sometimes called a "bracket expression"; thus bra and ket arise from the splitting of a bracket—bra(c)ket.

Another way in which kets may be defined is through eigenvalue problems. An eigenvector of the operator $\mathbf{R}$ corresponding to eigenvalue r_i is labeled by the eigenvalue, namely $|r_i\rangle$; thus:

$$\mathbf{R}|r_i\rangle = r_i|r_i\rangle. \tag{4}$$

A linear operator can be constructed by writing a bra and ket in the reverse of the order of expression 3; from the ket $|f_1\rangle$ and bra $\langle f_2|$ we have, for example, $|f_1\rangle \langle f_2|$. Formally adjoining this to the left of a ket $|g\rangle$ gives a linear operation on $|g\rangle$ in the sense

$$|f_1\rangle \langle f_2|g\rangle = (f_2, g)f_1, \tag{5}$$

the right-hand side being written in the older notation. Any sum of operators of this type may be seen to be also a linear operator.

An operator $|f\rangle \langle f|$, if $|f\rangle$ is normalized to unity, is a projection operator onto the axis whose unit vector is $|f\rangle$. In particular if the eigenvector $|r_i\rangle$ is normalized, $|r_i\rangle \langle r_i|$ projects onto the eigenaxis associated with the eigenvalue r_i.

Through this notation we can construct $\mathbf{E}(r)$ explicitly if $\mathbf{R}$ has only discrete eigenvalues, and its eigenvectors are thus a complete basis:

$$\mathbf{E}(r_1) = \sum_{r_i \leq r_1} |r_i\rangle \langle r_i|. \tag{6}$$

If used formally, this expression gives

$$\langle f| \, \mathbf{E}(r_1) \, |g\rangle = \langle f| \{ \sum_{r_i \leq r_1} |r_i\rangle \langle r_i| \} |g\rangle; \tag{7}$$

permuting integration with summation, we obtain

$$\langle f| \, \mathbf{E}(r_1) \, |g\rangle = \sum_{r_i \leq r_1} \langle f|r_i\rangle \langle r_i|g\rangle, \tag{8}$$

which is rigorously correct, given the definitions of $\langle f|r_i\rangle$, $\langle r_i|g\rangle$, and $|r_i\rangle$.

When **R** has a continuous eigenspectrum, its eigenfunctions are not elements of Hilbert space and thus cannot be fitted strictly into Dirac's scheme as so far stated. The scheme can, however, be extended along the following lines: Let $\phi(r, q)$ be an eigenfunction of **R** for eigenvalue r in the continuous spectrum satisfying

$$\mathbf{R}\phi(r, q) = r\phi(r, q), \tag{9}$$

and normalized as follows:

$$\int_{-\infty}^{\infty} \left| \int_{q_1}^{q_1 + \Delta q} \phi(r, q)\, dq \right|^2 dr = \Delta q. \tag{10}$$

Then we can define $\mathbf{E}(r)$ through

$$(f, \mathbf{E}(r_1)g) = \int_{-\infty}^{r_1} \left[\int \overline{f(q)}\phi(r, q)\, dq \right] dr \left[\overline{\phi(r, q)}g(q)\, dq \right]. \tag{11}$$

Dirac extends the bracket notation to the terms in the above integral. This may be regarded as a purely notational device:

$$\int \overline{f(q)}\phi(r, q)\, dq = \langle f|r\rangle,$$
$$\int \overline{\phi(r, q)}g(q)\, dq = \langle r|g\rangle, \tag{12}$$

whence

$$(f, \mathbf{E}(r_1)g) = \langle f| \mathbf{E}(r_1) |g\rangle$$
$$= \int_{-\infty}^{r_1} \langle f|r\rangle\, dr\, \langle r|g\rangle. \tag{13}$$

Dirac further proposes to "split the bracket" even when one of the members is not an element of Hilbert space, as in the preceding case of $\langle f|r\rangle$. The resulting symbols $|r\rangle$ and $\langle r|$ can be treated, purely formally, as elements of Hilbert space by procedures given in the cited book. If we also permute integrations in Equation 13, we obtain

$$\langle f| \mathbf{E}(r_1) |g\rangle = \langle f| \int_{-\infty}^{r_1} |r\rangle\, dr\, \langle r| \cdot |g\rangle, \tag{14}$$

whence we extract a formal expression for the projection operator,

$$\mathbf{E}(r_1) = \int_{-\infty}^{r} |r\rangle\, dr\, \langle r|. \tag{15}$$

Although this cannot be taken as a rigorous mathematical expression for $\mathbf{E}(r_1)$, the operator $\mathbf{E}(r_1)$ nonetheless exists, being defined by Equation 11. The expression 15 may be regarded as a handy condensation of the steps (the steps obtained by reversing the sequence of equations from 11 to 15) by which $(f, \mathbf{E}(r_1)g)$ is to be constructed.

It is sometimes useful to have the projection operator $\mathbf{E}(r)$ expressed in terms of its effect on operands that are functions of r itself. For this purpose, we use the step function $S_{r_1}(r)$ defined as follows:

$$S_{r_1}(r)f(r) = \begin{cases} f(r), & r \leq r_1, \\ 0, & r > r_1, \end{cases} \tag{16}$$

for all $f(r)$. We can then write

$$[\mathbf{E}(r_1)f](r) = S_{r_1}(r)f(r) \tag{17}$$

defining the element $\mathbf{E}(r)f$ resulting from the operation $\mathbf{E}(r)$ on f.

Another useful operator is the projection operator onto the manifold associated with an interval $(r, r + \Delta r)$, for which we introduce a special symbol

$$\mathbf{P}_{r, \Delta r} = \mathbf{E}(r + \Delta r) - \mathbf{E}(r). \tag{18}$$

If r_i is a discrete eigenvalue, the projection onto its eigenaxis will be denoted by

$$\mathbf{P}_{r_i} = \mathbf{E}(r_i) - \mathbf{E}(r_i - 0). \tag{19}$$

We now return to quantum mechanics. The postulates that remain to be stated define physical (that is, experimental) properties of systems. The first physical postulate is: The possible values obtainable on measurement of an observable are just the eigenvalues of its associated operator. (It is to be noted that we are bypassing the problem of the construction of the operator that goes with an observable; this is an important aspect of quantum mechanics in practice, but one which we need not go into.) When the eigenvalues are discrete, this yields the "hypothesis of quantization."

The other physical postulate defines the statistical content of quantum mechanics: Given a wave function ψ (an element of Hilbert space), normalized to unity, then the probability that a measurement of an observable having the operator $\mathbf{R}$ will yield a value equal to or less than r_1 is

$$F(r_1) = (\psi, \mathbf{E}(r_1)\psi). \tag{20}$$

Since this is the distribution function of r, expectation values of all functions of r can be computed from it, whatever $\mathbf{R}$ may be.

From Equation 20 it can be seen that it is only when ψ lies entirely within a manifold that is associated with only a single value of r (r then being a "degenerate" eigenvalue unless the manifold is one-dimensional) that the quantity R will be statistically undistributed. In no case can all possible observables be so. From this follows the "essentially" statistical character of quantum mechanics: Some statistical spread in the results of measurement will exist even when the system has the maximum amount of definition, corresponding to its being represented by an element of Hilbert space (ensembles of still less definition, not representable by single elements of Hilbert space, are also discussed in the theory, through the "statistical operator" of von Neumann [1932b, Chapters 4 and 5]). From the Bohr point of view, the wave function ψ still represents a single system—an individual. But as we have indicated in Section 1, we must regard ψ as representing rather an *ensemble* of individuals.

Our formulation of quantum mechanics will be completed by a statement of the manner in which the wave function varies in time; this will, by implication, yield the time variation of the distribution function $F(r)$. It is assumed that the physical properties of an isolated species of systems determine a hypermaximal Hermitian operator $\mathbf{H}$, called the Hamiltonian (because of a close formal relationship to the Hamiltonian function of classical dynamics), and that when (as we shall assume) the system is an isolated one, $\mathbf{H}$ is independent of the time. The time dependence of ψ is then given by the Schrödinger differential equation

$$i\hbar \frac{\partial \psi}{\partial t} = \mathbf{H}\psi, \tag{21}$$

($\hbar$ being Planck's constant h divided by 2π), from which the wave function at time t, ψ_t, can be expressed in terms of that at $t = t_0$:

$$\psi_t = \exp\left[-\frac{i}{\hbar}\mathbf{H}(t - t_0)\right]\psi_{t_0}. \tag{22}$$

4. Differential space of a quantum system

Let us consider first a hypermaximal Hermitian $\mathbf{Q}$ having a purely continuous spectrum with eigenvalues $-\infty < q < \infty$. It is possible to define a complex Brownian motion function $X_\alpha(q_1)$ such that for any interval $(q_1, q_1 + \varDelta q)$ the increment

$$\varDelta X_\alpha(q_1) = X_\alpha(q_1 + \varDelta q) - X_\alpha(q_1) \qquad (23)$$

has real and imaginary parts with Gaussian distribution, independent of one another and of the real or imaginary parts of $\varDelta X$ for any interval not overlapping $(q_1, q_1 + \varDelta q)$, and each with mean square $\varDelta q$.

Now let $f(q)$ be an arbitrary function belonging to $L_2(-\infty, \infty)$. We write

$$\langle f|\alpha\rangle = \int_{-\infty}^{\infty} \overline{f(q)} \, dX_\alpha(q); \qquad (24)$$

this defines the symbol $\langle f|\alpha\rangle$, the notation being suggested by the fact that the right-hand side has the form of an inner product of f with the formal derivative $X'_\alpha(q)$ of $X_\alpha(q)$ with respect to q. (Thus we take $|\alpha\rangle$ to be the "Hilbert space element"—actually of infinite length—corresponding to the *derivative* of $X_\alpha(q)$, not to $X_\alpha(q)$ itself.) The real and imaginary parts of $\langle f|\alpha\rangle$ have Gaussian distribution, each with mean square

$$E(\mathrm{Re}\,\langle f|\alpha\rangle)^2 = E(\mathrm{Im}\,\langle f|\alpha\rangle)^2$$
$$= \|f\|^2 = \int_{-\infty}^{\infty} |f(q)|^2 \, dq, \qquad (25)$$

and they are independent of one another and of any $\langle f_1|\alpha\rangle$ for which f_1 is orthogonal to f. The term $\varDelta X_\alpha(q_1)$ is a special case of the preceding example, with $f(q)$ the following function:

$$f(q) = \begin{cases} 1, & q_1 < q \leqq q_1 + \varDelta q, \\ 0, & \text{otherwise.} \end{cases} \qquad (26)$$

It follows from the definitions of the projection operators, definitions 17 and 18, that $\varDelta X_\alpha(q_1)$ is the total increment with respect to

q_1 of the projection of the element dX_α onto the manifold corresponding to the interval of eigenvalues $(q_1, q_1 + \Delta q)$:

$$\Delta X_\alpha(q_1) = \int_{-\infty}^{\infty} [\mathbf{P}_{q_1, \Delta q} \, dX_\alpha](q). \tag{27}$$

It is also necessary to formulate the transformation properties of the Brownian motion function, in order to define the measure function of differential space in any desired representation. Suppose (at first) $\mathbf{R}$ to have a continuous spectrum only, with eigenfunctions $\phi(r, q)$ (as in Section 3). The desired Brownian motion function will be expressed in terms of its increment. Its definition is suggested by the preceding formulation of $\Delta X_\alpha(q_1)$ as the increment of a projection of dX_α. The representation of the resolution of unity in 11 and the definition of the projection operator in 18 give, for an interval $(r_1, r_1 + \Delta r)$,

$$\begin{aligned}
\Delta X_\alpha(r) &= \int_{-\infty}^{\infty} [\mathbf{P}_{r_1, \Delta r} \, dX_\alpha](r) \\
&= \int_{r_1}^{r_1 + \Delta r_1} [\mathbf{P}_{r_1, \Delta r} \, dX_\alpha](r) \\
&= \int_{-\infty}^{\infty} \left[\int_{r_1}^{r_1 + \Delta r} \phi(r, q) \, dr \right] dX_\alpha(q). \tag{28}
\end{aligned}$$

The first and second forms are to be regarded as symbolic only; the last form is the correct one, since the interchange of order of integration is necessary to obtain a finite result.

The reader will please note that a somewhat novel functional notation is used in the above: $X_\alpha(r_1)$ is not the same function of r_1 as $X_\alpha(q)$ is of q; it is that function of r_1 obtained by transformation of $X_\alpha(q)$ according to the right-hand side of Equation 28. In this notation the letter used as argument tells what the function is, as well as serving as argument. Use of this convention makes for considerable simplicity in writing, and is based on the logic that the different representatives of the *same* element of a space may well be represented by the *same* symbol, the representation being indicated by the argument. Different values of the argument in the same representation may be indicated by subscripts, for example, r_1, r_2, etc.

From the discussion of the paragraph containing Equations 24 and 25, it follows that the distribution of $\Delta X_\alpha(r)$ has the same properties with respect to r as variable as $\Delta X_\alpha(q)$ was postulated to have with respect to q. This property will be referred to as *invariance of the distribution of* $\Delta X_\alpha(q)$ *with respect to unitary transformation.*

Finally, let us discuss the most general kind of eigenvalue spectrum that could concern us, namely one consisting of intervals and points —stretches (*strecken*) of continuous eigenvalues and discrete eigenvalues, respectively—distributed in any way on the infinite real line. If **R** is an operator with such a spectrum, then for a semiclosed interval $(r_1, r_1 + \Delta r]$ we will use in place of $\mathbf{P}_{r_1, \Delta r}$ of Equation 28 a projection operator which projects onto the sum of the Hilbert space manifolds corresponding to eigenvalues in this range. Let us denote eigenfunctions for discrete eigenvalues (normalized to unity) by $\phi_D(r_i, q)$ and those for continuous eigenvalues (normalized as in Section 3) by $\phi_c(r, q)$; then Equation 28 is replaced by

$$\Delta X_\alpha(r_1) = \int_{-\infty}^{\infty} \left[\sum_I \int_{I \cap (r_1, r_1 + \Delta r]} \phi_c(r, q)\, dr \right.$$
$$\left. + \sum_{r_1 < r \leq r_1 + \Delta r} \phi_D(r_i, q) \right] dX_\alpha(q), \qquad (29)$$

in which the first sum is over all intervals of eigenvalues that overlap $(r_1, r_1 + \Delta r]$ and the integral (as indicated by the symbols under the integral sign) is over the intersection of the interval I with $(r_1, r_1 + \Delta r]$. The sum in the second term is over the discrete eigenvalues r_i in $(r_1, r_1 + \Delta r]$.

The term $\Delta X_\alpha(r_1)$ will have the same properties of independence of real and imaginary parts mutually and with respect to those of projections onto nonoverlapping intervals of r as the foregoing functions. The variances of the real and imaginary parts will be equal to the total length of those parts of the intervals that are contained in $(r_1, r_1 + \Delta r]$ with the convention that spectral *lines* are in this respect to be considered as if they were eigenvalue intervals of length one. That is,

$$E[\text{Re } \Delta X_\alpha(r_1)]^2 = E[\text{Im } \Delta X_\alpha(r_1)]^2$$
$$= \sum_I m\{I \cap (r_1, r_1 + \Delta r]\} + n(r_1, \Delta r), \qquad (30)$$

where $m\{\ \}$ is the measure of an interval and $n(r_1, \Delta r)$ is the number of spectral lines in $(r_1, r_1 + \Delta r]$.

In order to sum up the situation in formal terms, we revert to a somewhat simpler situation. Take a continuous spectrum over the infinite line, with spectral lines superimposed (the more general situation of "stretches" can be handled in similar fashion): Formally, the distribution of $\Delta X_\alpha(r)$ may be summed up in terms of that of $dX_\alpha(r)$, the increment for the infinitesimal interval $(r, r + dr]$. The real and imaginary parts of $dX_\alpha(r)$ are Gaussian, mutually independent, independent of those for nonoverlapping intervals, and have variances equal to dr unless $(r, r + dr]$ contains a spectral line, in which case the variances are equal to one. In this connection, we can put

$$dX_\alpha(r_i) = \xi_\alpha(r_i), \tag{31}$$

when r_i is a discrete eigenvalue. Then ξ_α has Gaussian real and imaginary parts, independent of those of nonoverlapping intervals, and with unit variances. The preceding holds with respect to any representation, including q itself; thus, if Q has spectral lines, this makes it possible to interpret the q integration in Equation 29.

We can now generalize Equations 24 and 25, when R has discrete as well as continuous eigenvalues. Using the notation of definition 31, we write for a given $g(r)$,

$$\langle g|\alpha\rangle = \int_{-\infty}^{\infty} \overline{g(r)}\, dX_\alpha(r) + \sum g(r_i)\xi(r_i), \tag{32}$$

where the integration is over continuous r only, and the sum over discrete r only. The variable $\langle g|\alpha\rangle$ has real and imaginary parts with Gaussian distribution, mutually independent, and independent of any $\langle g_1|\alpha\rangle$ such that g_1 is orthogonal to g. The variances are

$$\begin{aligned}
E(\operatorname{Re}\langle g|\alpha\rangle)^2 &= E(\operatorname{Im}\langle g|\alpha\rangle)^2 \\
&= \|g\|^2 \\
&= \int_{-\infty}^{\infty} |g(r)|^2\, dr + \sum |g(r_i)|^2. \tag{33}
\end{aligned}$$

5. Assignment of hidden variables to points in differential space

It is implicit in the expression 20 for the distribution function of R,

$$F(r) = (\psi, \mathbf{E}(r)\psi), \tag{34}$$

that in the statistical theory of quantum mechanics, the projection operator and its associated linear manifold play a more fundamental role than the eigenvalues or sets thereof.

This is because according to the rule 20 the probability that R lies in a set is equal to the squared length of the projection of ψ onto the linear manifold defined by this set. Or, to symbolize, given an interval $(r, r + \Delta r]$, then from the projection operator $\mathbf{P}_{r,\Delta r}$ introduced at the end of Section 3, we obtain the probability for R to lie in $(r, r + \Delta r]$ as

$$(\psi, \mathbf{P}_{r,\Delta r}\psi). \tag{35}$$

For a sum of disjoint intervals, the sum of the corresponding projection operators—or, equivalently, the projection operator of the sum of the corresponding orthogonal manifolds—is used; the resulting probability is the sum of the probabilities for the component intervals. More complicated sets of points r than sums of intervals need not be considered.

Now, there is an infinite family of operators having the same resolution of unity as $\mathbf{R}$, namely all operators that commute with $\mathbf{R}$ or that have the form

$$\mathbf{Q}(\mathbf{R}) = \int q(r)\, d\mathbf{E}(r), \tag{36}$$

where $q(r)$ is an ordinary function of the real variable r. If the resolution of unity and the distribution function for $\mathbf{R}$ are known, the same are thereby known for $\mathbf{Q}(\mathbf{R})$. For suppose we require the probability that $\mathbf{Q}$ lie in a certain set. All that is required is to construct the set of inverse image points r, that is, such as map into values lying in the given set; then the desired probability is equal to the probability that r lies in the image set. Distribution problems in quantum mechanics are thus most compactly formulated in terms of resolutions of unity and the linear manifolds generated by them. We shall, therefore, in the following work put the emphasis on the idea of the realization of linear manifolds associated with intervals or sums of intervals of a variable, as being equivalent to the realization of the intervals or sums themselves and of more fundamental significance in the theory.

The differential-space theory of quantum systems is now characterizable as follows: It is a probability theory using as its sample space the differential space associated with the Hilbert space of the quantum-mechanical description of a system. It is postulated that the realized values of an observable must be eigenvalues of its associated hypermaximal operator. There is an algorithm (actually a choice of alternative algorithms) for the assignment of realized values of an observable to points in differential space, which amounts to formulating observables as functions of the random variable X_α; it can then be proved that the distribution function of any observable as so constructed agrees with that of quantum mechanics. The algorithm will, as implied in the previous paragraph, be stated in terms of the realization of linear manifolds associated with the resolution of unity of the operator, this being equivalent to the assignment of realized values of the operator or any function thereof. Finally, the dynamical properties of the individual systems represented by the points α can be defined so that their time development is deterministic to any desired precision, consistently with the probabilities as defined, and thus in agreement with quantum mechanics as far as all distributions (which is all that quantum mechanics defines) are concerned.

In order to formulate the algorithms, we introduce certain definitions, as follows: Given a linear manifold $\mathfrak{M}$; the projection operator onto $\mathfrak{M}$ to be called $\mathbf{P}_{\mathfrak{M}}$. The projection of the wave function ψ onto $\mathfrak{M}$ will be denoted by $\psi_{\mathfrak{M}}$:

$$\psi_{\mathfrak{M}} = \mathbf{P}_{\mathfrak{M}}\psi. \tag{37}$$

For example: Let ψ be given as a function of q, and let the eigenfunctions of $\mathbf{R}$ be $\phi(r, q)$ (as in Section 3). Then the representative of ψ is[8]

$$\psi(r) = \int_{-\infty}^{\infty} \phi(r, q)\psi(q) \, dq. \tag{38}$$

[8] The same convention as to functional notation is used here as in Equation 28.

If $\mathfrak{M}$ is the manifold associated with the interval of eigenvalues $(r_1, r_1 + \Delta r]$, then $\psi_\mathfrak{M}(r)$ is given by

$$\psi_\mathfrak{M}(r) = \begin{cases} \psi(r), & r_1 < r \leq r_1 + \Delta r, \\ 0, & \text{otherwise.} \end{cases} \tag{39}$$

We have for the probability of realizing the interval Δr, for a ψ normalized to unity, from Equation 20

$$\text{Prob}\,\{r \text{ in } (r_1, r_1 + \Delta r]\} = (\psi, \mathbf{P}_\mathfrak{M}\psi) = \|\mathbf{P}_\mathfrak{M}\psi\|^2$$
$$= \int_{r_1}^{r_1 + \Delta r} |\psi(r)|^2 \, dr + \sum_{r_1 < r_i \leq r_1 + \Delta r} |\psi(r_i)|^2, \tag{40}$$

the last being the earliest and most elementary way of formulating the statistical hypothesis of quantum mechanics.

It should be apparent in the foregoing equation that the single interval $(r_1, r_1 + \Delta r_1]$ could be replaced by any sum of disjoint intervals. The space $\mathfrak{M}$ consists then of the sum of the corresponding manifolds.

Although we have given only a one-dimensional example, the method of using manifolds to characterize ranges of physical observables is perfectly general, applying to systems of any number of spatial dimensions, in any representation (that is, momentum, energy, etc., may be used instead of position), and to further variables such as spin as well; and these variables may be continuous or quantized. We have not yet given details about the construction of the Brownian motion functions in the multidimensional case. This will be done later; for the moment let it suffice to say that this can be done in perfect analogy to the one-dimensional case.

The Dichotomic Algorithm. Let $\mathfrak{M}_1$ be the manifold associated with a sum of intervals of eigenvalues of an observable R, and $\mathfrak{M}_2$ its orthogonal complement in $\mathfrak{H}_\infty$,

$$\mathfrak{M}_1 \oplus \mathfrak{M}_2 = \mathfrak{H}_\infty. \tag{41}$$

Let there be given an ensemble of systems of maximum experimental definition, represented by the wave function ψ. Projecting ψ onto $\mathfrak{M}_1$ and $\mathfrak{M}_2$, form the functionals

$$\langle \psi_{\mathfrak{M}_1} | \alpha \rangle \quad \text{and} \quad \langle \psi_{\mathfrak{M}_2} | \alpha \rangle. \tag{42}$$

According to Equations 32 and 33 and accompanying discussion, these are independently distributed, with variances equal to $\|\psi_{\mathfrak{M}_1}\|^2$ and $\|\psi_{\mathfrak{M}_2}\|^2$ for the real and imaginary parts of the respective functionals. These variances are also the respective probabilities for the experimental realization of the manifolds $\mathfrak{M}_1$ and $\mathfrak{M}_2$ according to quantum mechanics.

One of the two manifolds is now assigned to α, namely that for which the associated functional is the larger. This is the desired correspondence, or at least the first stage in a sequence of such correspondences, by which manifolds of r are made into random functions of α:

$$\mathfrak{M}(\alpha) = \mathfrak{M}_i, \tag{43}$$

where i is such that

$$|\langle \psi_{\mathfrak{M}_i} | \alpha \rangle| = \max \{ |\langle \psi_{\mathfrak{M}_1} | \alpha \rangle|, \ |\langle \psi_{\mathfrak{M}_2} | \alpha \rangle| \}. \tag{44}$$

It can now be shown that the random function $\mathfrak{M}(\alpha)$ has the correct distribution, that is, the probabilities calculated therefrom for the realization of manifolds $\mathfrak{M}_1$ and $\mathfrak{M}_2$ agree with the quantum-mechanical hypothesis, Equation 40. The probability that, say, $|\langle \psi_{\mathfrak{M}_1} | \alpha \rangle| > |\langle \psi_{\mathfrak{M}_2} | \alpha \rangle|$ is (with the known characterization of these two quantities) equal to

$$\frac{1}{(2\pi)^2} \frac{1}{a_1 a_2} \iiiint\limits_{\xi_1^2 + \eta_1^2 > \xi_2^2 + \eta_2^2} \exp\left[-\frac{\xi_1^2 + \eta_1^2}{2a_1} - \frac{\xi_2^2 + \eta_2^2}{2a_2} \right] d\xi_1 \, d\eta_1 \, d\xi_2 \, d\eta_2 \tag{45}$$

(in which a_i, where $i = 1, 2$, stands for $\|\psi_{\mathfrak{M}_i}\|^2$ and the quantities ξ_i, η_i correspond to the real and imaginary parts, respectively, of $\langle \psi_{\mathfrak{M}_i} | \alpha \rangle$). The inequality under the integral sign defines the region of integration. That is, the integral is to be carried out over the four-dimensional region satisfying this inequality. The integral is easily evaluated in polar coordinates (Wiener and Siegel [1953]). The result is a_1, as desired.

In their first publication, Wiener and Siegel contemplated that the first dichotomy of the Hilbert space might be the first of a sequence of dichotomies by which a particular α is assigned to an arbitrarily small manifold (i.e., to a manifold $\mathfrak{M}(\alpha)$ for which $\|\psi_{\mathfrak{M}(\alpha)}\|$ is as small as desired). The value of $\|\psi_{\mathfrak{M}(\alpha)}\|$ at the final step would then be chosen to correspond to the precision of the experimental observations.

In order to examine this supposition, let us suppose that on the first dichotomy α is assigned to $\mathfrak{M}_1$, and decompose $\mathfrak{M}_1$ into two parts $\mathfrak{M}_{11}$ and $\mathfrak{M}_{12}$. We follow the pattern of the first dichotomy by assigning α to $\mathfrak{M}_{1i}$, where i is such that

$$|\langle\psi_{\mathfrak{M}_{1i}}|\alpha\rangle| = \max\{|\langle\psi_{\mathfrak{M}_{11}}|\alpha\rangle|, |\langle\psi_{\mathfrak{M}_{12}}|\alpha\rangle|\}. \tag{46a}$$

Denote the set of all points α of differential space which are assigned to $\mathfrak{M}_1$ on the first dichotomy by S_1, and the set of points assigned to $\mathfrak{M}_{1i}$ on the second by S_{1i}. Now if P is the probability measure on differential space, agreement with ordinary quantum theory requires

$$P(S_1 \cap S_{1i}) = \|\psi_{\mathfrak{M}_{1i}}\|^2; \qquad i = 1, 2, \cdots. \tag{46b}$$

However, as the calculation in the Appendix shows, Equation 46b does not hold in general.[9] This was not appreciated by Wiener and Siegel at the time of their 1953 paper.

The reason for this failure of the second dichotomy to give Equation 46b is that the first dichotomy, more precisely, the imposition of Equation 46a, destroys the independence of the differential-space variables $\langle\psi_{\mathfrak{M}_{11}}|\alpha\rangle$, $\langle\psi_{\mathfrak{M}_{12}}|\alpha\rangle$. Their joint distribution when subjected to this condition becomes proportional to

$$\exp\left[-\frac{1}{2}\left(\frac{|\langle\psi_{\mathfrak{M}_{11}}|\alpha\rangle|^2}{\|\psi_{\mathfrak{M}_{11}}\|^2} + \frac{|\langle\psi_{\mathfrak{M}_{12}}|\alpha\rangle|^2}{\|\psi_{\mathfrak{M}_{12}}\|^2}\right)\right]$$
$$\times \left[1 - \exp\left(-\frac{1}{2}\frac{|\langle\psi_{\mathfrak{M}_1}|\alpha\rangle|^2}{\|\psi_{\mathfrak{M}_2}\|^2}\right)\right] \tag{46c}$$

(this expression follows from the integral A.10 of the Appendix), which in view of the second factor in the bracketed term is not a product of functions of $\langle\psi_{\mathfrak{M}_{11}}|\alpha\rangle$ and $\langle\psi_{\mathfrak{M}_{12}}|\alpha\rangle$ separately.

To be sure, Equation 46b *is* guaranteed if $\langle\psi_{\mathfrak{M}_{11}}|\alpha\rangle$ and $\langle\psi_{\mathfrak{M}_{12}}|\alpha\rangle$ have equal variance; this is also shown in the Appendix. In general,

[9] This was first noted by Dr. Robert L. Warnock. We are indebted to Dr. Warnock for the analysis of the next two paragraphs, and for his kindness in contributing the Appendix.

however, such a restriction is not imposable in the case of observables with discrete eigenvalue spectra. Moreover, even when imposable it destroys the possibility of making the partition of the eigenvalue spectrum conform to the nature of the measurement. The *polychotomic algorithm*, to be described in the next paragraph, provides a way of avoiding these difficulties.

The Polychotomic Algorithm. This algorithm carries out in a single step any desired subdivision of $\mathfrak{H}_\infty$ into constituent submanifolds. Corresponding to a set of n such, $\{\mathfrak{M}_i \mid i = 1, 2, \cdots, n\}$, a mapping $\mathfrak{M}(\alpha)$ is constructed according to the criterion

$$\mathfrak{M}(\alpha) = \mathfrak{M}_i, \qquad (47a)$$

where i is such that

$$\frac{|\langle \psi_{\mathfrak{M}_i} | \alpha \rangle|^2}{\| \psi_{\mathfrak{M}_i} \|^4} = \min \left\{ \frac{|\langle \psi_{\mathfrak{M}_1} | \alpha \rangle|^2}{\| \psi_{\mathfrak{M}_1} \|^4}, \frac{|\langle \psi_{\mathfrak{M}_2} | \alpha \rangle|^2}{\| \psi_{\mathfrak{M}_2} \|^4}, \cdots, \frac{|\langle \psi_{\mathfrak{M}_n} | \alpha \rangle|^2}{\| \psi_{\mathfrak{M}_n} \|^4} \right\}. \qquad (47b)$$

Since the quantities

$$\frac{\text{Re} \langle \psi_{\mathfrak{M}_k} | \alpha \rangle}{\| \psi_{\mathfrak{M}_k} \|} \qquad \text{and} \qquad \frac{\text{Im} \langle \psi_{\mathfrak{M}_k} | \alpha \rangle}{\| \psi_{\mathfrak{M}_k} \|}$$

are all mutually independent and Gaussian with variance 1, the probability that Equation 47a is satisfied with, say, $i = 1$ is

$$\int_{\left\{ \frac{\xi_1^2 + \eta_1^2}{a_1} < \frac{\xi_2^2 + \eta_2^2}{a_2}, \frac{\xi_3^2 + \eta_3^2}{a_3}, \cdots, \frac{\xi_n^2 + \eta_n^2}{a_n} \right\}} \cdots (2n\text{-fold}) \cdots \int \exp \left[-\tfrac{1}{2} \sum_{j=1}^{n} (\xi_j^2 + \eta_j^2) \right] \prod_{j=1}^{n} \frac{d\xi_j \, d\eta_j}{(2\pi)^n} \qquad (48)$$

(see Wiener and Siegel [1955] for detailed calculation) which can be evaluated, to give the result a_1, agreeing with quantum mechanics.

In carrying out either the dichotomic or the polychotomic construction there will, of course, be a set of points α for which the minima called for are multivalued. For these points the function $\mathfrak{M}(\alpha)$ may be simply left undefined, as they constitute a set of zero measure. We can thus obtain distributions of all observables agreeing with quantum mechanics even for arbitrary partitions of their eigenvalue spectra, and we thereby escape the restrictions on the theory that would follow from the restricted scope of the dichotomic theorem.

The general possibility of carrying out successive polychotomies, envisaged in Wiener and Siegel (1953), is ruled out in the same way that successive dichotomies are ruled out. That is, one cannot expect to apply the polychotomic theorem as it stands to submanifolds of the manifold which satisfies a set of polychotomic inequalities of the type of Equation 47b. It is not necessary to construct a formal proof. It is sufficient to note (and this has been verified by the calculation in the Appendix) that the imposition of an inequality on the modulus of a sum of complex random variables necessarily engenders correlations among them, because some sets of arguments will favor the inequality, while others will work against it, and the outcome will depend on the values of the arguments *relative* to one another. Thus the polychotomic inequality will result in the destruction of the mutual independence of the components of the differential-space vector within the manifold $\mathfrak{M}_i$ of statement 47a, leading to a correlated distribution analogous to that exhibited in expression 46c.

This means, in particular, that the treatment of the quantum theory of measurement proposed by Siegel and Wiener [1956] is invalid, because they wrongly assumed the continued independence of differential-space components after satisfaction of a polychotomy. (See also in this respect, Ochs [1964, p. 50].) The incorporation of the observer's recognition of the result of the measurement process into the differential-space theory thus remains troublesome, as it indeed is also in standard quantum mechanics.

Note on distributions of two or more observables. Actual work in quantum mechanics usually involves wave functions of several variables; for example, four variables, one (continuous) for each of the three spatial dimensions and one (discrete) for spin, are commonly required. All these are observables; time, on which the wave function also depends, in general, must be considered as a "parameter" rather than as an observable, and is not one of the dimensions underlying the Hilbert space used.

Ordinary quantum mechanics is easily modified for this purpose. In constructing a resolution of unity in the Hilbert space, all variables must be used. This is most readily formulated by combining them into a vector, each of whose orthogonal components is one of the individual variables. Some of the components (such as spin) may have a discrete range of values, others a continuous range (for instance, spatial variables). The formulas of Section 3 may be retained with a simple change of notation: r is replaced by a vector $\vec{r}$, $\mathbf{R}$ by $\vec{\mathbf{R}}$ (meaning a vector operator whose components are scalar operators), and the resolution of unity is given by $\mathbf{E}(\vec{r})$ (not a vector). The components of $\vec{r}$ and $\vec{\mathbf{R}}$ will be denoted by the symbol without the arrow and with a superscript indicating the particular component, e.g., $r^{(k)}$ is the kth component of $\vec{r}$. The individual eigenvalue problems are assumed soluble:

$$\mathbf{R}^{(k)} = \int_{-\infty}^{\infty} r^{(k)} \, d\mathbf{E}(r^{(k)}), \tag{49}$$

where, in the spirit of the notation introduction in Section 4 and explained just below Equation 28, $\mathbf{E}(r^{(k)})$ is by definition the projection operator for the resolution of unity in the $\mathbf{R}^{(k)}$ representation, and $\mathbf{E}(r^{(k)})$ and $\mathbf{E}(r^{(l)})$ for any l different from k are thus in general different functions of their arguments. Then the resolution of unity for the vector operator $\vec{\mathbf{R}}$ is given by a projection operator in the product space, the direct product of the projection operators for the components

$$\mathbf{E}(\vec{r}) = \mathbf{E}(r^{(1)})\mathbf{E}(r^{(2)}), \cdots, \mathbf{E}(r^{(n)}), \tag{49}$$

if $\vec{r}$ has the n components $r^{(1)}, \cdots, r^{(n)}$. The vector operator $\vec{\mathbf{R}}$ itself is given by

$$\vec{\mathbf{R}} = \int \vec{r} \, \overrightarrow{d\mathbf{E}(\vec{r})}, \tag{50}$$

where the integral is over the n variables, and $\overrightarrow{d\mathbf{E}(\vec{r})}$ is not a vector, but a common conventional notation for the product $d\mathbf{E}(r^{(1)})$ $\times \, d\mathbf{E}(r^{(2)}) \cdots d\mathbf{E}(r^{(n)})$.

With wave functions depending on n variables, Equation 50 replaces Equation 1. Similarly, all formulas of Section 3 may be retained with scalar operators on eigenvalues retained by vector ones, and single integrals by multiple ones. Where one-dimensional eigenvalues are integrated over intervals or sets of intervals, n-dimensional eigenvalues are integrated over sets of points in n-dimensional space.

The Brownian motion function $X_\alpha(\vec{r})$ is now defined by a procedure analogous to that of Section 4, in which an interval of r is replaced by a set of intervals, one for each component of $\vec{r}$. Such a set defines a set of points in the n-dimensional space consisting of all those whose coordinates are eigenvalues lying in the corresponding interval. (As with the one-dimensional case, parts of each interval may be devoid of eigenvalues, corresponding to a spectrum consisting of points or intervals.) Such a set of intervals may be denoted by $\Delta\vec{r}$, a vector whose components are the individual component intervals (a particular set of axes in the $\vec{r}$ space is implicitly understood). Associated with the point $\vec{r}$ and with the "increment" $\Delta\vec{r}$, we have a random function $\Delta X(\vec{r})$. The increment notation in $\Delta X(\vec{r})$ is also only conventional; $\Delta X(\vec{r})$ is really a random function of the set $\Delta\vec{r}$. Its real and imaginary parts are mutually independent, and independent of those for any nonoverlapping $\Delta\vec{r}$. The variances of the real and imaginary parts are both equal to the n-dimensional volume of eigenvalue space within $\Delta\vec{r}$, computed as the volume of an n-dimensional parallelepiped with edges parallel to the coordinate axes; the edge length of this parallelepiped with respect to any particular eigenvalue coordinate (say the ith) is to be taken equal to the total length of the eigenvalue intervals of $r^{(i)}$ contained in $\Delta r^{(i)}$, counting spectral lines (as in Section 4) as intervals of length one.

As the n-dimensional generalization of Equation 24 we have

$$\langle f|\alpha\rangle = \int_{-\infty}^{\infty} \cdots (n\text{-fold}) \cdots \int_{-\infty}^{\infty} f(\vec{r})\, dX_\alpha(\vec{r}).$$

All the equations following 24 now have n-dimensional generalizations obtained by merely replacing scalar quantities by vectors, using the generalized projection operator and the above properties of $\Delta X_\alpha(\vec{r})$.

The statistical postulate of quantum mechanics (Equation 20) is also unchanged in form when an n-dimensional $\vec{r}$ and the correct projection operator (Equation 49) are used.

Thus the discussion earlier in the present section involving intervals of eigenvalues is extended to the n-dimensional case by using n-dimensional sets in the eigenvalue space, and the corresponding projection operators. In particular, the dichotomic and polychotomic algorithms apply, dividing the eigenvalue space into two or more nonoverlapping sets, these replacing the intervals of the one-dimensional case; such a decomposition then generates the linear manifolds in $\mathfrak{H}_\infty$ to which values of α are assigned.

6. Change with time of the hidden variables

We now introduce a fundamental new postulate concerning the association between systems and their representative points in differential space under the passage of time: *If a system (belonging to a species having Hamiltonian operator* H, *with* H *independent of the time) is associated at a given instant with the point* $|\alpha\rangle$, *then its representative point at time t later is*

$$e^{-(i/\hbar)\,\mathbf{H}t}\,|\alpha\rangle. \tag{51}$$

Thus representative points change with time exactly like wave functions (see Equation 22). Expression 51 symbolizes the result of applying to the point α the differential-space transformation that is adjoint to $\exp[-(i/\hbar)\mathbf{H}t]$, as defined in Wiener and Akutowicz [1957]. The correspondence between the Hilbert-space transformation and its adjoint is such that we can symbolize the latter by the former.

It will be noted that the transformation 51 satisfies an important necessary condition if the set of points of differential space is to represent a statistical ensemble of conceivably real systems: The measure (probability) of a set of points (systems) is conserved in time under the transformation. This is because the transformation $\exp[-(i/\hbar)\mathbf{H}t]$ is unitary and, as noted in Section 4, measure in differential space is preserved under the adjoint of a unitary transformation.

The transformation 51 is of fundamental importance for the physical interpretation of our theory, because it determines the behavior in time of the individual systems of our ensembles, systems which, if we allow ourselves to speculate, might some day be isolated experimentally as individuals rather than as inseparable members of an ensemble. Hence we shall now derive the laws of variation of the physical observables of one of these systems.

First, we note that the unitary operator of 51, which we denote by $\mathbf{T}_t$,

$$\mathbf{T}_t = e^{-(i/\hbar)\,\mathbf{H}t}, \tag{52}$$

generates from an operator $\mathbf{R}$ an infinite family of operators

$$\mathbf{R}(t) = \mathbf{T}_t^{-1}\mathbf{R}\mathbf{T}_t$$

by similarity transformation, with values of t ranging from $-\infty$ to $+\infty$. As is well known, all the operators $\mathbf{R}(t)$ have the same eigenvalue spectrum.

Imagine now that either the dichotomic or polychotomic algorithm is applied to all the hypermaximal operators of a system, subject to the restriction that the successive dichotomic partitions of the eigenvalue spectrum, or the single polychotomic partition, in the respective schemes, are carried out in exactly the same way for all operators belonging to the same family under the above similarity transformation. This means, according to the more fundamental language in terms of which we decompose the Hilbert space into manifolds defined by projection operators of the form $\mathbf{P}_{r,\Delta r}$ (we confine the discussion to scalar operators for simplicity—the generalization to vector operators is readily made), that where a set of projection operators $\mathbf{P}_1, \mathbf{P}_2, \cdots$ defines the decomposition used with respect to an operator $\mathbf{R}$, the projection operators for $\mathbf{R}_t$ would be $\mathbf{T}_t^{-1}\mathbf{P}_1\mathbf{T}_t$, $\mathbf{T}_t^{-1}\mathbf{P}_2\mathbf{T}_t$, $\cdots$. However, these projection operators will be seen to correspond to the same sets of eigenvalues of $\mathbf{R}_t$ as the untransformed ones to sets of eigenvalues of $\mathbf{R}$.

We now note the following: The algorithms of our theory concern relationships among the quantities α, ψ and the operators $\mathbf{P}_1, \mathbf{P}_2, \cdots$ from which the manifolds denoted by $\mathfrak{M}_i$ were constructed. The result of the algorithms is an association of manifolds—or, equivalently, of projection operators—with points α, determined by ψ. Now suppose one of the algorithms to be applied to the projection operators of $\mathbf{R}_t$, namely $\mathbf{T}_t^{-1}\mathbf{P}_1\mathbf{T}_t$, $\mathbf{T}_t^{-1}\mathbf{P}_2\mathbf{T}_t$, $\cdots$, using the point α and wave function ψ. The result is a certain assignment of projection operators, or eigenvalue intervals of $\mathbf{R}_t$, to α. Next, suppose the same algorithm to be applied to the projection operators of $\mathbf{R}$, namely $\mathbf{P}_1, \mathbf{P}_2, \cdots$, using the future values of α and ψ, namely $\mathbf{T}_t\alpha$ and $\mathbf{T}_t\psi$. The quantities here used have the same relationships among one another as those of the prior case, in the sense that the quantities in one case are all unitary transforms of those of the other; the norm of $\mathbf{P}_1 \cdot \mathbf{T}_t\psi$ is the same as that of $\mathbf{T}_t^{-1}\mathbf{P}_1\mathbf{T}_t \cdot \psi$, the inner product between $\mathbf{P}_1 \cdot \mathbf{T}_t\psi$ and $\mathbf{T}_t\alpha$ is the same as that between $\mathbf{T}_t^{-1}\mathbf{P}_1\mathbf{T}_t \cdot \psi$ and α, and so forth. Thus at the later time, the system (represented by $\mathbf{T}_t\alpha$) is assigned to a range of $\mathbf{R}$ which is the same as that range of $\mathbf{R}_t$ to which it was assigned earlier. Thus we may say that the value of $\mathbf{R}$ (as closely as we may wish to determine it) at a future time t is equal to that of $\mathbf{R}_t$ at the present time. (Note that $\mathbf{R}_t$, despite the tag t, is in principle an observable quantity at the *present* time, since it is a hypermaximal operator.) Hence the future values of all dynamical variables are determined by their present ones, as closely as we wish, and the individual system may therefore fairly be said to behave in a deterministic fashion.

We shall now give a qualitative interpretation, according to our theory, of the experiment of Section 2, in which an ensemble of particles passes through a pair of slits and emerges on the other side with a density following the optical diffraction pattern that would be produced by the two slits—a density distribution depending, in particular, on the separation of the two slits. Our picture would be of an ensemble of individual particles endowed with an infinity of dynamical variables in addition to the classical ones of position and momentum. (An infinity of dynamical variables governing a single particle need not seem inappropriate when we compare with the case of the motion of a charged particle in an electric field, since in full generality the electric field also has an infinite number of degrees of freedom.) The totality of these variables expresses the influence of the diaphragm, with its two slits, on the behavior of each particle. In fact, the entire future history of the particle is contained in its present, in the form of variables whose values are the future values of other variables. Thus the particle will go through a definite slit, one or the other, following a path determined by its set of dynamical variables. (It will not, of course, behave like a particle subjected to ordinary forces.) While each particle follows its precise preordained path, the ensemble as a whole will emerge past the diaphragm so constituted as to yield a diffraction pattern in its spatial density.

Admittedly such a picture is unduly elaborate, and inferior to quantum mechanics for all present computational purposes. Apart from the emotional satisfaction it may furnish to those who enjoy seeing anomalous facts reduced to familiar categories (categories familiar in the sense of previously existing, as compared to the ad hoc, though certainly brilliant, explanations of the Copenhagen school), its value will only be proved if it can be shown ultimately to have the ability to explain some otherwise inexplicable experimental effect. But of this, more later.

At this point it is appropriate to deal with an objection raised by Schwartz in the seminar notes of Friedrichs, Shapiro, *et al.* [1957, Chapter 15]. It is to be emphasized that the hidden variables of the individual systems are the values of the observables (that is, of all functionally independent **R**) attached to the point α, and not the coordinates of α itself. The latter must be considered to serve entirely in the capacity of a mathematical aid rather than as a physically meaningful entity; a scaffolding, so to speak, for erecting the ensemble in terms of values of observables, which once this is accomplished, serves no further purpose than that of an arbitrary tag by means of which measures of sets may be calculated. If this is not borne in mind, the description previously given of particle motion in terms of hidden parameters would appear to be quite inadmissible, on the following grounds: The values of observables attached to a point α will vary from one ensemble to another, since they depend on the wave function. But the statistical point of view allows only the probability of a set of observables to be determined by the statistical situation (by the ensemble), and not the values of the observables themselves. That is, a statistical ensemble consists of a set of possible, precise situations individually quite unassociated with probability implications, on which a probability measure is then imposed without changing the individual situations. It is Schwartz's claim that the present theory transgresses this requirement of a statistical theory.

When it is realized that the "individual situation" in our case is not really determined by α, we have the key to the solution of this difficulty. Consider two ensembles with different ψ's: a pair of systems, one from each ensemble, having the same α, do not constitute the same "individual situation" at all; but a pair, having the same values of all observables **R**, do. Where then does the difficulty arise? In the following: That we have not, in our scheme, quite literally followed the statistical procedure mentioned above, namely of defining our probability measure over a set of physical observables. Instead, we have defined it over a set of arbitrary tags α, which are correlated with the physical observables, to be sure, but differently so with different ensembles.

Nothing prevents our imagining our ensembles expressed in terms of the more direct formulation, as follows: Start with an ensemble to which one of the algorithms has been applied, down to reasonably small intervals of each and every observable, so that we can say that a "value" (actually, a small range) of each observable has been assigned to each system. Now construct a space having as orthogonal axes the functionally independent observables—strictly speaking, not all of them, since they form a continuum, but a large finite set of them in some sense sufficiently representative of the entire set of observables. The ensemble originally in differential space may be transferred to this space, assigning to any region of the latter a measure which we know how to obtain from the prior application of the algorithm. This space, not differential space, is the counterpart of the "phase space," or space of physical observables, of classical statistical physics, and we shall therefore so term it. *Any* ensemble may be transferred to this same space. In it, each individual system has attached to it by definition a set of physical observables quite independent of the wave function, or the probability measure. The probability measure, on the other hand, does vary as it should from one ensemble to another. Moreover, we have seen how to follow a system α from one instant to the next, through the hypothesis 51, and with preservation of measure. By successively computing the observables attached to this point with the algorithm, we can define the motion of a system point through phase space also. We have seen that the change in time of the physical observables of a phase point is entirely determined by the values of the observables at this point. Hence there is no question of the wave function influencing the temporal behavior of the phase point. Since measure is preserved in time for a set of systems obeying expression 51, it must follow that measure is preserved in phase space too.

If phase space is regarded as primary, differential space may be regarded figuratively as a bed of Procrustes, to whose fixed measure every ensemble is forced to conform, for the sake of convenience in constructing ensembles by means of the algorithms. This reduction to a common measure function could actually be done with any kind of ensemble, including those of classical physics, yielding ensembles in which the observable values attached to systems seem to (but do not) depend on the probability distribution.

7. Von Neumann's theorem on hidden variables

If the differential space formulation of quantum theory is correct, then all physical experiments analyzed to date have, for all intents and purposes, been carried out on ensembles constructible from some wave function ψ through one of our two algorithms.[10] The peculiar implication of a hidden-variable theory, as such, is that ensembles not so constructible exist and can in principle be discovered in nature. A hidden-variable theory, such as ours, is thus a metatheory within which situations following the strict quantum-mechanical laws can be described, but other situations diverging from these laws have a place as well.

An extreme instance of a situation disagreeing with quantum mechanics would be an ensemble consisting of a single point of our phase space. Such an ensemble is at least possible in principle, even though we might in practice be willing to settle for much less spectacular confirmations of our theory. It would present unique values of all physical observables. But this raises the question of the relationship of the differential-space theory to von Neumann's well-known proof [1932b, pp. 157–173] of the impossibility, subject to certain conditions, of "dispersionless" (*streuungslos*) ensembles. This proof is commonly taken to demonstrate the impossibility of

[10] Or on a linear combination of such ensembles, represented by von Neumann's statistical operator (also referred to in Section 4). For purposes of our discussion, these are not to be considered as differing essentially from those constructed from a single ψ—the "pure states."

constructing a successful hidden-parameter theory: "dispersionless" means having zero variance for all observables, and this is indeed the case with our one-point ensemble.

There is no contradiction between our point ensemble and von Neumann's proof, however, because one of the conditions of his proof is not satisfied by our one-point ensemble. This is his postulate B', which says that the quantum-mechanical expectation value of a sum of observables is the sum of the quantum-mechanical expectation values of the individual observables:

$$E_{qu}(R + S + \cdots) = E_{qu}(R) + E_{qu}(S) + \cdots. \qquad (53)$$

Note that we distinguish the quantum-mechanical expectation value by the subscript "qu," since it is presumably intrinsically different from that of ordinary probability theory. (In our theory it is not, of course, but corresponds simply to an expectation value over a restricted class of ensembles.)

The condition 53 cannot be true of our one-point ensemble. A counterexample to it can be seen as follows: The expectation values of observables in the one-point ensemble are, of course, the unique values of the observables belonging to the single phase point. Thus we would have to have just

$$(R + S + \cdots)_{\text{realized}} = R_{\text{realized}} + S_{\text{realized}} + \cdots, \qquad (54)$$

where the subscript denotes the value of the operator at the phase point. But the possible "realized" values of an observable in our theory are, as stated in Section 5, restricted to the eigenvalues of its associated observable. The left-hand side of Equation 54 is thus an eigenvalue of $R + S + \cdots$, and the right-hand side, a sum of eigenvalues of R, of S, $\cdots$. The eigenvalues of a sum of operators are often not at all related to the sums of the eigenvalues of the individual operators in a way that would accord with Equation 54. For example, the eigenvalues of the square of the coordinate operator in one dimension, x^2, are continuous from $-\infty$ to $+\infty$, and the same is true of the square of the momentum operator p^2. On the other hand, the eigenvalues of the sum $p^2 + x^2$ (this is the Hamiltonian of a "harmonic oscillator"—see any textbook of quantum mechanics) are a discrete set only. Thus, in ensembles ranging over a continuum of values of p^2 and of x^2 individually, there will necessarily occur values of α (and over sets of nonzero measure) for which the sum of these values does not equal any of the quantized eigenvalues of $p^2 + x^2$.

Von Neumann's imposition of his condition B' must presumably have been motivated by its apparent physical reasonableness. Hence a certain onus is placed on us for rejecting it, and we should make some attempt to justify our doing so. At the very least, we can claim that there is no need for insisting on this condition. Granted that all experimental data so far support the quantum-mechanical ensembles, and that these satisfy condition B', we assert (a) that our theory does include these, and (b) that these data cannot be extrapolated to impose condition B' on exceptional ensembles, since our formalism shows that quantum-mechanical ensembles, satisfying the condition, and exceptional ensembles, not satisfying it, are logically compatible with each other.

Speculatively, the argument may be pushed somewhat further. If point ensembles are considered to have an objective meaning, as being, for example, idealized limiting cases of less exceptional ensembles that might some day be realized experimentally, then even the spirit of present-day quantum mechanics seems to deny the validity of condition B' as applied to them. Consider an operator represented as a sum of two other operators,

$$\mathbf{Q} = \mathbf{R} + \mathbf{S}. \tag{55}$$

Such a representation is quite arbitrary, in the sense that $\mathbf{R}$ may be *any* operator, provided $\mathbf{S} = \mathbf{Q} - \mathbf{R}$. In particular, $\mathbf{R}$ may not commute with $\mathbf{Q}$. In this context, we now ask whether it is reasonable at all to expect that the values of the observables for $\mathbf{Q}$, $\mathbf{R}$, and $\mathbf{S}$, namely Q, R, and S, satisfy the same relation, that is,

$$Q = R + S \tag{56}$$

for the point ensemble. One may well expect otherwise, since if $\mathbf{R}$ does not commute with $\mathbf{Q}$, quantum mechanics (in one of the most essential postulates of the Copenhagen school) claims that a measurement of each will necessarily alter the value of the other in an unpredictable fashion (this is the "uncontrollable disturbance" discussed in Section 1). Although we must deny that this is eternally true, since it would contradict the eventual realizability of exceptional ensembles, we may take it as evidence that the observable Q is to some extent independent of R and of S, if $\mathbf{R}$ and $\mathbf{S}$ do not commute with $\mathbf{Q}$.

8. A historical analogy

The basic features of the differential-space theory have now been presented. It must be admitted that it is an extraordinarily difficult theory for computational purposes, and even if it were worked out considerably beyond its present state would still be inferior to quantum mechanics in cases to which the latter applies exactly. And we are further prepared to admit that the purely intellectual satisfaction of reducing quantum mechanics to the terms of ordinary probability theory, and of thereby making it more compatible with ordinary intuition, does not confer any truly substantial merit on the proposed theory when so much simplicity is lost thereby. Of what value is the theory then, can we justifiably plead for its further development and that it be not consigned to oblivion?

In considering this question, we would like to invoke an interesting historical precedent, namely that of the one-time struggle between statistical mechanics and thermodynamics. Although these two branches of physics live compatibly side by side nowadays, it was not always so. Toward the latter part of the nineteenth century, Boltzmann's contention that statistical mechanics provides a fundamental explanation of the laws of thermodynamics was sharply challenged, both technically (for a review of this controversy, see P. and T. Ehrenfest [1911]) and philosophically (in particular by Mach [1896]).

The following single quotation is a vivid statement of the philosophical arguments of Mach [1896, p. 364].

The mechanical conception of the Second Law through the distinction between *ordered* and *unordered* motion, through the establishment of a parallel between the increase of entropy and the increase of ordered motion at the expense of disordered, seems quite artificial. If one realizes that a real analogy of the entropy *increase* in a purely mechanical system consisting of absolutely elastic atoms does not exist, one can hardly help thinking that a violation of the Second Law—and without the help of any demon—would have to be possible if such a mechanical system were the real seat of thermal processes. I agree here with F. Wald completely, when he says, "In my opinion the roots of this (entropy) law lie much deeper, and if success were achieved in bringing about agreement between the molecular hypothesis and the entropy law, this would be fortunate for the hypothesis, but not for the entropy law."

The echoes of the scientific-philosophic battles of the nineteenth century over the foundations of physics are faint indeed to our ears now, their intensity largely absorbed in their passage through the corridors of time. We are their beneficiaries, and the relative clarity of our present views is a product of these tests and struggles. The views of Mach bear upon the controversy over the choice between a molecular-mechanical foundation for thermodynamics and one free of such elements and therefore more purely postulational. In a sense, each side, as we see it now, won a partial victory. Thermodynamics is indeed now established as a discipline that is autonomous in principle, needing no help from mechanics in the logical sense. But in practice the statistical-mechanical point of view permeates the expressed thoughts of physicists in every area to which it is at all relevant. The achievements of kinetic theory and statistical mechanics since the active days of Mach have been such that the idea that molecular motions are the source of the laws of thermodynamics has gained universal acceptance. This is all the more striking, in that the logical argument by which the laws are to be deduced from these motions is still incomplete. It is simply that the phenomenal success of the molecular hypothesis in modern physics, compared to its relatively primitive stage of development in the nineteenth century, makes any attempt to study temperature-dependent phenomena without its aid seem like a pedantic eccentricity.

Viewed retrospectively then, the attempt to deny the molecular origin of thermodynamic laws failed because the facts made it obsolete. Yet philosophical considerations of simplicity and postulational economy seemed to support it at the time. We would like to argue that these considerations are directly applicable to the situation in regard to *hidden parameters* in quantum mechanics at the present time. In order to emphasize how apt this parallel is, we wish to cite the use by Mach [1896, p. 363] of virtually this same term in characterizing kinetic theory. In admitting the attractiveness and fruitfulness of mechanical pictures for the scientist's intuition, he attempts to relegate them to an auxiliary position in the following words: "The freedom gained by assuming *hidden, invisible motions* ('unsichtbare verborgene Bewegungen') is fundamentally no greater than in the case of Black's assumption of a latent heat" [our italics].

It is, in fact, striking that the philosophical school with which Mach is identified, namely *logical positivism*, is also the one credited with furnishing the inspiration for the current interpretation of quantum mechanics. In view of the stupendous successes of quantum mechanics, we would hardly wish to argue against the value of a program of postulational purification for constructing a useful scientific theory. But one is also bound to consider the limitations of any philosophy, however successful; *to the case of thermodynamics versus statistical mechanics, which demonstrated limitations to the logical-positivist point of view in the field of thermal phenomena, we propose as a possible parallel the case of quantum mechanics versus hidden-parameter quantum theories.*

Just as the hidden-parameter point of view in thermodynamics was vindicated only by its empirical success, we would expect that the same in quantum systems must wait for the decision as to its merit upon empirical verifications that are yet to come. But by the same token, no argument at present possible can disprove its validity (we are thinking, of course, of the general point of view and not of any special formulation of it). In particular, citations of the success of quantum mechanics are irrelevant, just as the undoubted success of the purified thermodynamics did not prevent statistical mechanics from reaching its present prominence. Nor on the other hand does the eventual potential vindicability of the hidden-parameter point of view imply that the quantum-mechanical point of view will not continue to be successful; the future achievements of the former are, we think, to be sought in areas quite different from those in which quantum mechanics is currently applied, different in ways that are necessarily difficult to anticipate at present.

If we wish to speculate on the way in which deviations from quantum mechanics might be found, we may draw from the thermodynamics-statistical mechanics case as a parallel. Thermodynamics arises from statistical mechanics in the following way: Ensembles of nonequilibrium systems, with their macroscopic observables statistically distributed in arbitrary fashion, tend toward stationary "equilibrium" ensembles (in a way approximately understood, although not yet rigorously derived), in which the macroscopic observables have almost unique values, that is, have almost zero statistical dispersion. We may call these the "normal" values of the observables. Thermodynamics is the study of the functional relationships between the normal values of the observables, as they are changed by altering the conditions of the ensembles. Thus it is concerned with certain special distributions, namely the equilibrium ones, but due to the special circumstance of almost zero dispersion it is able to ignore the question of distribution altogether, and take the form of an entirely nonstatistical theory.

We wish now to propose that the distributions of quantum mechanics are the counterpart of the dispersionless ensembles of thermodynamics. Quantum-mechanical ensembles are not dispersionless, to be sure, nor are they constant in time, but they may, in a future experimentally validated theory of quantum systems, be singled out in a sense similar to that in which dispersionless distributions are singled out in the thermodynamics-statistical mechanics. The analogy would then suggest that quantum ensembles deviating from the quantum-mechanical distribution might be produced by a sudden, severe disturbance of a system, and that the deviations would then decay in time (Bohm [1953] has suggested such a possibility in terms of his hidden-parameter theory); the decay time would have to be quite short for such deviations to have eluded observation till now. If this is so, it would be appropriate to seek for hidden-parameter effects in the form of small deviations from the quantum-mechanical distribution, by rapid observation after a disturbance.

Technique of Computing Joint Probabilities and Remarks on the Dichotomic Algorithm

By Prof. Robert L. Warnock
Department of Physics
Illinois Institute of Technology

The sets in differential space determined by the dichotomic or polychotomic algorithm are specified by inequalities of the type:

$$|\langle \psi_1 | \alpha \rangle|^2 \geq a |\langle \psi_2 | \alpha \rangle|^2; \qquad a > 0. \tag{A.1}$$

Measures of intersections of such sets are interesting, since they correspond to joint probabilities of two or more observables having values in specified intervals. For example, we may obtain the joint probability that $q_1 \leq q \leq q_2$, $p_1 \leq p \leq p_2$, where q and p are the position and momentum of a particle, respectively. This is a quantity of a sort not contemplated in standard quantum theory, so it would be desirable to compute it explicitly in terms of the wave function. Such computations are difficult in general because of the complicated geometry involved in intersections of the sets. However, a great deal of simplification comes about if we represent inequality A.1 by means of the step function, and then use a sort of Fourier representation of the step function. That is, let

$$\theta(x) = \begin{cases} 1, & x \geq 0, \\ 0, & x < 0, \end{cases}$$

and insert a factor $\theta(|\langle \psi_1 | \alpha \rangle|^2 - a |\langle \psi_2 | \alpha \rangle|^2)$ in the integration which is to be restricted by inequality A.1. Then employ the following identity:

$$\int_{-\infty}^{\infty} \theta(x)f(x)\,dx = \lim_{\epsilon \to 0+} \frac{1}{2\pi i} \int_{-\infty}^{\infty} \frac{d\lambda}{\lambda - i\epsilon} \int_{-\infty}^{\infty} e^{i\lambda x} f(x)\,dx. \quad \text{(A.2)}$$

We first prove Equation A.2 under the conditions:

$$\int_{-\infty}^{\infty} |f(x)|\,dx < \infty; \qquad f(x) \in C, \quad \text{(A.3)}$$

which are much stronger than necessary, but adequate for our purposes. We define

$$\phi_1(\lambda) + \phi_2(\lambda) = \int_{-\infty}^{0} e^{i\lambda x} f(x)\,dx + \int_{0}^{\infty} e^{i\lambda x} f(x)\,dx, \quad \text{(A.4)}$$

where $\phi_1(\lambda)$ is analytic for Im $\lambda < 0$, while $\phi_2(\lambda)$ is analytic for Im $\lambda > 0$. By contour integration,

$$\frac{1}{2\pi i} \int_{-\infty}^{\infty} \frac{\phi_1(\lambda)\,d\lambda}{\lambda - i\epsilon} = 0,$$

and $\qquad\qquad\qquad\qquad\qquad\qquad\qquad\qquad\qquad\qquad$ (A.5)

$$\frac{1}{2\pi i} \int_{-\infty}^{\infty} \frac{\phi_2(\lambda)\,d\lambda}{\lambda - i\epsilon} = \phi_2(i\epsilon).$$

Equation A.2 follows. The contribution of the closing contour at infinity in the lower (upper) half-plane is zero. That is, one uses the Parseval relation

$$\int_{-\infty}^{\infty} |\phi_1(u + iv)|^2\,du = 2\pi \int_{-\infty}^{0} |f(x)|^2 \, e^{-2vx}\,dx < M; \qquad v \le 0,$$

$$\text{(A.6)}$$

to make a simple estimate of the integral over a rectangular contour receding to infinity.

Equation A.2 may be applied to compute a joint probability involved in the dichotomic algorithm; viz., the quantity $P(S_1 \cap S_{1i})$ of Equation 46b, Chapter 5. We use the following notations for the random variables of expression 42, Chapter 5, as follows:

$$\langle \psi_{\mathfrak{M}_1} | \alpha \rangle = x_1 + iy_1 = z_1, \qquad \langle \psi_{\mathfrak{M}_2} | \alpha \rangle = x_2 + iy_2 = z_2,$$
$$\langle \psi_{\mathfrak{M}_{11}} | \alpha \rangle = X_1 + iY_1 = Z_1, \qquad \langle \psi_{\mathfrak{M}_{12}} | \alpha \rangle = X_2 + iY_2 = Z_2,$$
$$|z_i|^2 = r_i^2, \qquad\qquad |Z_i|^2 = R_i^2. \qquad \text{(A.7)}$$

Only three of the variables are independent; in fact, $z_1 = Z_1 + Z_2$. Since $\psi_{\mathfrak{M}_2}, \psi_{\mathfrak{M}_{11}}, \psi_{\mathfrak{M}_{12}}$ form an orthogonal set, z_2, Z_1, Z_2 are independent. Their variances are denoted as

$$a_2 = \|\psi_{\mathfrak{M}_2}\|^2, \qquad a_{11} = \|\psi_{\mathfrak{M}_{11}}\|^2, \qquad a_{12} = \|\psi_{\mathfrak{M}_{12}}\|^2. \qquad \text{(A.8)}$$

The measure to be calculated is

$$P(S_1 \cap S_{11})$$
$$= (2\pi)^{-3}(a_2 a_{11} a_{12})^{-1} \int\limits_{R_1^2 > R_2^2} dX_1\, dY_1\, dX_2\, dY_2 \exp\left[-\frac{1}{2}\left(\frac{R_1^2}{a_{11}} + \frac{R_2^2}{a_{12}}\right)\right]$$
$$\times \int\limits_{|Z_1 + Z_2|^2 > r_2^2} dx_2\, dy_2 \exp\left[-\frac{1}{2}\frac{r_2^2}{a_2}\right]. \qquad \text{(A.9)}$$

Evaluation of the x_2, y_2 integral by polar coordinates leaves the fourfold integral

$$(2\pi)^{-2}(a_{11} a_{12})^{-1} \int\limits_{R_1^2 > R_2^2} dX_1\, dY_1\, dX_2\, dY_2 \exp\left[-\frac{1}{2}\left(\frac{R_1^2}{a_{11}} + \frac{R_2^2}{a_{12}}\right)\right]$$
$$\times \left[1 - \exp\left(-\frac{1}{2a_2}|Z_1 + Z_2|^2\right)\right]. \qquad \text{(A.10)}$$

Again by polar coordinates the first term is found to be equal to $a_{11}/(a_{11} + a_{12})$. To handle the second term we introduce Equation A.2 to obtain

$$\lim_{\epsilon \to 0+} -(2\pi i\, a_{11} a_{12})^{-1}(2\pi)^{-2} \int_{-\infty}^{\infty} d\lambda(\lambda - i\epsilon)^{-1}$$
$$\times \int_{-\infty}^{\infty} dX_1\, dY_1\, dX_2\, dY_2 \exp\left[i\lambda(R_1^2 - R_2^2)\right]$$
$$\times \exp\left[-\frac{1}{2}\left(\frac{R_1^2}{a_{11}} + \frac{R_2^2}{a_{12}} + \frac{|Z_1 + Z_2|^2}{a_2}\right)\right]. \qquad \text{(A.11)}$$

To justify this step, change to polar coordinates $R_1^2, \theta_1, R_2^2, \theta_2$ in the inner integral and then make the change of variable $R_1^2, R_2^2 \to x = R_1^2 - R_2^2, y = R_1^2 + R_2^2$. By leaving the x integration to be done last, one finds that the function $f(x)$ of Equation A.3 vanishes exponentially at infinity, so expression A.11 is indeed correct. After returning to Cartesian coordinates we carry out the inner integrations by the standard formula:

$$(2\pi)^{-n/2} \int_{-\infty}^{\infty} \prod_{i=1}^{n} dx_i \exp\left[-\frac{1}{2}\sum_{i,j=1}^{n} x_i A_{ij} x_j\right] = (\det A)^{-\frac{1}{2}}, \quad (A.12)$$

which holds since the real part of A is positive definite. The determinant of A happens to be a perfect square (fortunately, this is typical of all joint probability calculations of the theory). An evaluation of the λ integral by the method of residues leads easily to the result

$$P(S_1 \cap S_{11}) = \frac{a_{11}}{a_1} + \frac{2a_{11}a_{12}}{\beta(a_{11} - a_{12} - \beta)}, \quad (A.13)$$

$$\beta = \left[(a_{11} - a_{12})^2 + \frac{4a_{11}a_{12}}{a_2}\right]^{\frac{1}{2}}. \quad (A.14)$$

By the interchange $a_{11} \leftrightarrow a_{12}$, $P(S_1 \cap S_{12})$ is obtained from A.13. As a check one verifies that

$$P(S_1 \cap S_{11}) + P(S_1 \cap S_{12}) = P(S_1) = a_1.$$

In general, Equation A.13 does not give the result a_{11} required for agreement with quantum mechanics. If $a_{11} = a_{12} = a_1/2$, however, one can readily show that it does give such agreement. One then obtains

$$P(S_1 \cap S_{11}) = P(S_1 \cap S_{12}) = a_1/2.$$

This can also be shown directly from the original differential-space distribution, by the following symmetry argument. When

$$\|\psi_{\mathfrak{M}_{11}}\| = \|\psi_{\mathfrak{M}_{12}}\|, \quad (A.15)$$

the variables $\langle\psi_{\mathfrak{M}_{11}}|\alpha\rangle$ and $\langle\psi_{\mathfrak{M}_{12}}|\alpha\rangle$ are statistically indistinguishable. They have equal variance by Equation A.15, and their covariances with the variables which define S_1 are also equal. The latter statement is true because of the relations

$$\langle\psi_{\mathfrak{M}_1}|\psi_{\mathfrak{M}_{11}}\rangle = \langle\psi_{\mathfrak{M}_1}|\psi_{\mathfrak{M}_{12}}\rangle = \|\psi_{\mathfrak{M}_{11}}\|^2, \quad (A.16)$$

and

$$\langle\psi_{\mathfrak{M}_2}|\psi_{\mathfrak{M}_{11}}\rangle = \langle\psi_{\mathfrak{M}_2}|\psi_{\mathfrak{M}_{12}}\rangle = 0.$$

Hence,

$$P(S_1 \cap S_{11}) = P(S_1 \cap S_{12}). \quad (A.17)$$

But $S_{11} \cup S_{12}$ is the whole space, and $P(S_{11} \cap S_{12}) = 0$. Therefore

$$P(S_1 \cap S_{11}) + P(S_1 \cap S_{12}) = P(S_1) = \|\psi_{\mathfrak{M}_1}\|^2, \qquad \text{(A.18)}$$

and, by Equation A.17,

$$P(S_1 \cap S_{1i}) = \|\psi_{\mathfrak{M}_{1i}}\|^2 = \tfrac{1}{2}\|\psi_{\mathfrak{M}_1}\|^2; \qquad i = 1, 2. \qquad \text{(A.19)}$$

It is clear that successive dichotomies will continue to yield agreement with standard quantum theory as long as we require the two random variables of each dichotomy to have equal variance. In that case, their covariances with all variables involved in previous dichotomies will also be equal, and the symmetry argument applies.

We may ask whether any relation other than $a_{11} = a_{12}$ will yield $P(S_1 \cap S_{11}) = a_{11}, P(S_1 \cap S_{12}) = a_{12}$ (i.e., agreement with quantum theory). Putting $a_{11}/a_{12} = \alpha$, we find that $P(S_1 \cap S_{11}) = a_{11}$ requires a certain eighth-degree polynomial in α to vanish. Therefore, α is not arbitrary as had been claimed in earlier work.

Bibliography

AKUTOWICZ, EDWIN J.
1957 "On an explicit formula in linear least squares prediction," *Math. Scand.* **5**, 261–266.

BACHELIER, L.
1900 "Théorie de la spéculation," *Ann. Sci. École Norm. Sup.* s.3, **17**, 21–86.
1901 "Théorie mathématique du jeu," *Ann. École Norm. Sup.* s.3, **18**, 143–210.
1912 *Calcul des Probabilités*, Gauthier-Villars, Paris.
1913 "Les probabilités cinématiques et dynamiques," *Ann. École Norm. Sup.* s.3, **30**, 77–119.
1937 *Les Lois des Grands Nombres du Calcul des Probabilités*, Gauthier-Villars, Paris.

BILLINGSLEY, P. P.
1956 "The invariance principle for dependent random variables," *Trans. Amer. Math. Soc.* **83**, 250–268.

BIRKHOFF, GEORGE D.
1931 "Proof of the ergodic theorem," *Proc. Nat. Acad. Sci. U.S.A.* **17**, 656–660.

BOCHNER, SALOMON
1947 "Stochastic processes," *Ann. Math.* **48**, 1014–1061.

BOHM, DAVID
1952 "Suggested interpretation of the quantum theory in terms of 'hidden' variables: I, II," *Phys. Rev.* **85**, 166–179, 180–193.
1953 "Proof that probability density approaches $|\psi|^2$ in causal interpretation of quantum theory," *Phys. Rev.* **89**, 458–466.
1957 *Causality and Chance in Modern Physics*, Routledge & Kegan Paul, London. D. Van Nostrand, Princeton.

BOHR, NIELS
1935 "Can quantum-mechanical description of reality be considered complete?" *Phys. Rev.* **48**, 696–702.

CAMERON, R. H., and W. T. MARTIN
1944 "The Wiener measure of Hilbert neighborhoods in the space of real continuous functions," *J. Math. & Phys.* **23**, 195–209.
1945a "Wiener integrals under a general class of linear transformations," *Trans. Amer. Math. Soc.* **58**, 184–219.

1945*b* "Evaluation of various Wiener integrals by use of certain Sturm-Liouville differential equations," *Bull. Amer. Math. Soc.* **51**, 73–90.

1947 "The orthogonal development of nonlinear functionals in series of Fourier-Hermite functionals," *Ann. Math.* **48**, 385–392.

CHANDRASEKHAR, S.
1943 "Stochastic problems in physics and astronomy," *Rev. Mod. Phys.* **15**, 1–89. Also included in Wax [1954].

CRAMÉR, HARALD
1940 "On the theory of stationary random processes," *Ann. Math.* **41**, 215–230.

DARLING, D. A., and A. J. F. SIEGERT
1953 "The first passage problem for a continuous Markov process," *Ann. Math. Stat.* **24**, 624–639.

DERMAN, C.
1954 "Ergodic properties of the Brownian motion process," *Proc. Nat. Acad. Sci. U.S.A.* **40**, 1155–1158.

DIRAC, P. A. M.
1958 *Principles of Quantum Mechanics*, Fourth Edition, The Clarendon Press, Oxford.

DONSKER, MONROE D.
1951 "An invariance principle for certain probability limit theorems," *Memoirs Amer. Math. Soc.* No. 6, 12 pp.

DOOB, J. L.
1942 "The Brownian movement and stochastic equations," *Ann. Math.* **43**, 351–369. Also included in Wax [1954].

1953 *Stochastic Processes*, John Wiley & Sons, New York.

EHRENFEST, P. and T.
1911 "Begriffliche Grundlagen der statistischen Auffassung in der Mechanik," *Enzyklopädie der mathematischen Wissenschaften*, IV 2, II, Heft 6, B. G. Teubner, Leipzig. Also in *Collected Scientific Papers by Paul Ehrenfest*, edited by Martin J. Klein, North-Holland, Amsterdam, and Interscience, New York, 1959. Also trans. into English by Michael J. Moravcsik, Cornell University Press, Ithaca, 1959.

EINSTEIN, ALBERT
1905 "Über die von der molekularkinetischen Theorie der Wärme geforderte Bewegung von in ruhenden Flüssigkeiten suspendierten Teilchen," *Ann. Physik* **17**, 549–560.

1906 "Zur Theorie der Brownschen Bewegung," *Ann. Physik* **19**, 371–381.

1936 "Physik und Realität," *J. Franklin Inst.* **221**, 313–347; trans. into English, 349–382.

EINSTEIN, A., B. PODOLSKY, and N. ROSEN
1935 "Can quantum-mechanical description of reality be considered complete?" *Phys. Rev.* **47**, 777–780.

ERDÖS, PAUL, and MARK KAC
1946 "On certain limit theorems of the theory of probability," *Bull. Amer. Math. Soc.* **52**, 292–302.
1947 "On the number of positive sums of independent random variables," *Bull. Amer. Math. Soc.* **53**, 1011–1020.

FELLER, WILLIAM
1951 "The asymptotic distribution of the range of sums of independent random variables," *Ann. Math. Stat.* **22**, 427–432.

FEYERABEND, P. K.
1962 "Problems of microphysics," *Frontiers of Science and Philosophy*, edited by Robert G. Colodny, University of Pittsburgh Press, Chap. 6.

FEYNMAN, RICHARD P.
1951 "The concept of probability in quantum mechanics," *Proceedings of the Second Berkeley Symposium on Mathematical Statistics and Probability*, University of California Press, Berkeley, pp. 533–541.

FORTET, ROBERT
1943 "Les fonctions aléatoires du type de Markoff associées à certaines équations linéaires aux dérivées partielles du type parabolique," *J. de Math.* **22**, 177–243.
1949 "Quelques travaux récents sur le mouvement Brownien," *Ann. Inst. H. Poincaré* **11**, 175–225.
1958 "Recent advances in probability theory," *Some Aspects of Analysis and Probability*, by I. Kaplansky, Marshall Hall, Edwin Hewitt, and Robert Fortet, *Surveys in Applied Math.*, Vol. 4, John Wiley & Sons, New York, pp. 171–243.

FRIEDRICHS, K. O., H. N. SHAPIRO, *et al.*
1957 *Integration of Functionals*, New York University, Institute of Mathematical Sciences.

FURRY, W. H.
1936 "Note on the quantum-mechanical theory of measurement," *Phys. Rev.* **49**, 393–399.

FÜRTH, R.
1917 "Einige Untersuchungen über Brownsche Bewegung an einem Einzelteilchen," *Ann. Physik* **53**, 177–213.

HALMOS, PAUL R.
1956 *Lectures in Ergodic Theory*, The Mathematical Society of Japan, Tokyo.

HOPF, EBERHARD
  1937 "Ergodentheorie," *Ergeb. Math.* **5**, No. 2. Reprinted by Chelsea Publishing Co., New York, 1948.

ITO, KIYOSHI
  1944 "Stochastic integral," *Proc. Imp. Acad. Tokyo* **20**, 519–524.

KAC, MARK
  1946 "On the average of a certain Wiener functional and a related limit theorem in calculus of probability," *Trans. Amer. Math. Soc.* **59**, 401–414.
  1947 "Random walk and the theory of Brownian motion," *Amer. Math. Monthly* **54**, 369–391. Also included in Wax [1954].
  1949 "On distributions of certain Wiener functionals," *Trans. Amer. Math. Soc.* **65**, 1–13.
  1951 "On some connections between probability theory and differential and integral equations," *Proceedings of the Second Berkeley Symposium on Mathematical Statistics and Probability*, University of California Press, Berkeley, pp. 189–215.
  1959 *Probability and Related Topics in Physical Sciences: Lectures in Applied Mathematics*, **1**, Interscience Publishers, London.

KAC, MARK, and A. J. F. SIEGERT
  1947 "On the theory of noise in radio receivers with square law detectors," *J. App. Phys.* **18**, 383–397.

KALLIANPUR, G., and HERBERT ROBBINS
  1953 "Ergodic property of the Brownian motion process," *Proc. Nat. Acad. Sci. U.S.A.* **39**, 525–533.

KOLMOGOROV, A. N.
  1933 "Grundbegriffe der Wahrscheinlichkeitsrechnung," *Ergeb. Math.* **2**, No. 3. Trans. into English by Nathan Morrison, Chelsea Publishing Co., New York (1950).
  1941a "Stationary sequences in Hilbert space" (Russian), *Bull. Mat. Moscov. Univ.* **2**, No. 6, 40 pp.
  1941b "Interpolation und Extrapolation von stationären zufälligen Folgen" (Russian, German summary), *Izv. Akad. Nauk S.S.S.R., Ser. Mat.* **5**, 3–14.

LANING, J. H., and R. H. BATTIN
  1956 *Random Processes in Automatic Control*, McGraw-Hill, New York

LÉVY, PAUL
  1939 "Sur certains processus stochastiques homogènes," *Compositio Math.* **7**, 283–339.
  1940 "Le mouvement Brownien plan," *Amer. J. Math.* **62**, 487–550.
  1948 *Processus Stochastiques et Mouvement Brownien*, Gauthier-Villars, Paris.

1953 "Random functions: General theory with special reference to Laplacian random functions," *University of California Publications in Statistics* 1, No. 12, 331–390.

LOÈVE, MICHEL
1955 *Probability Theory. Foundations. Random Sequences*, D. van Nostrand, New York. See also Third Edition (1963).

LOUD, W. S.
1954 "A nonexceptional element of Wiener space," *Proc. Amer. Math. Soc.* 5, 940–941.

MACH, ERNST
1896 *Prinzipien der Wärmelehre*, J. A. Barth, Leipzig.
1923 *Prinzipien der Wärmelehre*, Fourth Edition, J. A. Barth, Leipzig.

MARK, A. M.
1949 "Some probability limit theorems," *Bull. Amer. Math. Soc.* 55, 885–900.

MOYAL, J. E.
1949 "Quantum mechanics as a statistical theory," *Proc. Cambridge Phil. Soc.* 45, 99.

VON NEUMANN, JOHANN
1932a "Proof of the quasi-ergodic hypothesis," *Proc. Nat. Acad. Sci. U.S.A.* 18, 70–82.
1932b *Mathematische Grundlagen der Quantenmechanik*, Julius Springer, Berlin. Reprinted by Dover Publications, New York, 1943. Trans. into English by Robert T. Beyer, Princeton University Press, Princeton (1955).
1950 *Functional Operators*, Vol. II: *The Geometry of Orthogonal Spaces*, Annals of Mathematics Studies, No. 22, Princeton University Press, Princeton.

OCHS, WILHELM
1964 *Über die Wiener-Siegelsche Formulierung der Quantentheorie*, unpublished thesis, Frankfurt-am-Main.

PALEY, R. E. A. C., NORBERT WIENER, and ANTONI ZYGMUND
1933 "Notes on random functions," *Math. Z.* 37, 647–668.

PALEY, R. E. A. C., and NORBERT WIENER
1934 *Fourier Transforms in the Complex Domain*, Amer. Math. Soc. Colloq. Pub. 19, American Mathematical Society, New York.

PERRIN, JEAN B.
1910 *Brownian Movement and Molecular Reality*, Trans. F. Soddy, Taylor & Francis, London.

RADEMACHER, HANS
1922 "Einige Sätze über Reihen von allgemeinen Orthogonalfunktionen," *Math. Ann.* 87, 112–138.

RANKIN, BAYARD
1960 "Computable probability spaces," *Acta Math.* **103**, 89–122.
1965 "Quantum mechanical time," *J. Math. Phys.* **6**, 1057–1071.
1966*a* "The history of probability and the changing concept of the individual," *J. History of Ideas*, **27**, No. 4.
1966*b* "Quantized energy distributions for the harmonic oscillator," *Phys. Rev.* **141**, 1223–1230.

RICE, S. O.
1944 "Mathematical analysis of random noise," *Bell System Tech. J.* **23**, 282–332. Also included in Wax [1954].
1945 "Mathematical analysis of random noise," *Bell System Tech. J.* **24**, 46–156. Also included in Wax [1954].

SCHIFF, LEONARD
1955 *Quantum Mechanics*, McGraw-Hill, New York.

SCHILPP, PAUL ARTHUR, Ed.
1949 *Albert Einstein, Philosopher-Scientist*, The Library of Living Philosophers, Inc., Evanston.

SIEGEL, ARMAND, and NORBERT WIENER
1956 "'Theory of measurement' in differential-space quantum theory," *Phys. Rev.* **101**, 429–432.

VON SMOLUCHOWSKI, MARYAN
1916 "Drei Vorträge über Diffusion, Brownsche Molekularbewegung und Koagulation von Kolloidteilchen," *Physic. Z.* **17**, 557–571, 585–599.
1918 "Über den Begriff des Zufalls und den Ursprung der Wahrscheinlichkeitsgesetze in der Physik," *Naturwiss.* **6**, 253–263.

SZEGÖ, GABOR
1939 *Orthogonal Polynomials*, Amer. Math. Soc. Colloq. Pub. **23**, American Mathematical Society, New York.

UHLENBECK, G. E., and L. S. ORNSTEIN
1930 "On the theory of the Brownian motion," *Phys. Rev.* **36**, 823–841. Also included in Wax [1954].

UHLENBECK, G. E., and J. L. LAWSON
1950 *Threshold Signals*, M.I.T. Rad. Lab. Series, Vol. 24, McGraw-Hill, New York.

WANG, M. C., and G. E. UHLENBECK
1945 "On the theory of Brownian motion, II," *Rev. Mod. Phys.* **17**, 323–342. Also included in Wax [1954].

WAX, NELSON
1954 *Selected Papers on Noise and Stochastic Processes*, Dover Publications, New York.

WEYL, HERMANN
1916 "Über die Gleichverteilung von Zahlen mod. Eins," *Math. Ann.* **77**, 313–352.

1931   *The Theory of Groups and Quantum Mechanics*, Trans. from Second (German) Edition by H. P. Robertson, Methuen & Co., Ltd., London.

WIENER, NORBERT

1923   "Differential space," *J. Math. & Phys.* **58**, 131–174. Also included in Wiener [1965].

1930   "Generalized harmonic analysis," *Acta Math.* **55**, 117–258. Also included in Wiener [1965].

1939   "The ergodic theorem," *Duke Math. J.* **5**, 1–18. Also included in Wiener [1965].

1949*a* *Extrapolation, Interpolation, and Smoothing of Stationary Time Series*, The Technology Press and John Wiley, New York. Reprinted by The M.I.T. Press, Cambridge, Massachusetts (1964).

1949*b* "Sur la théorie de la prévision statistique et du filtrage des ondes," *Analyse Harmonique, Colloques Internationaux du CNRS*, No. 15, 67–74, Centre National de la Recherche Scientifique, Paris.

1955   "On the factorization of matrices," *Comm. Math. Helv.* **29**, 97–111.

1965   *Selected Papers of Norbert Wiener*, with contributions by Y. W. Lee, N. Levinson, W. T. Martin, Society for Industrial and Applied Mathematics and The M.I.T. Press, Cambridge, Massachusetts.

WIENER, NORBERT, and EDWIN J. AKUTOWICZ

1957   "The definition and ergodic properties of the stochastic adjoint of a unitary transformation," *Rend. Circ. Mat. Palermo* **6**, 1–13.

1959   "A factorization of positive Hermitian matrices," *J. Math. & Mech.* **8**, 111–120. Also included in Wiener [1965].

WIENER, NORBERT, and P. MASANI

1957   "The prediction theory of multivariate stochastic processes," *Acta Math.* **98**, 111–150.

1958   "The prediction theory of multivariate processes, II," *Acta Math.* **99**, 93–137.

WIENER, NORBERT, and ARMAND SIEGEL

1953   "A new form for the statistical postulate of quantum mechanics," *Phys. Rev.* **91**, 1551–1560.

1955   "The differential-space theory of quantum systems," *Nuovo Cimento* **2**, Ser. X (*Supp. No.* 4) 982–1003.

WIGNER, EUGENE

1932   "On the quantum correction for thermodynamic equilibrium," *Phys. Rev.* **40**, 749.

WOLD, HERMAN
   1938   *A Study in the Analysis of Stationary Time Series*, Almqvist
          & Wiksell, Uppsala.
ZASUHIN, V.
   1941   "On the theory of multidimensional stationary random
          processes," *Dokl. Akad. Nauk*. S.S.S.R. **33**, 435–437.
ZYGMUND, ANTONI
   1935   *Trigonometrical Series*, Monografje Matematyczne, War-
          sawa-Lwów.

Index

169